Lean Six Sigma Green Belt Course Slide Book

Lean Six Sigma Green Belt Course Slide Book

Mary McShane-Vaughn

University Training Partners Publishing

University Training Partners Publishing

Folsom, CA

https://courses.6sigma.university

Copyright © 2024 by University Training Partners Publishing. All rights reserved. Except as permitted by US Copyright Law, no part of this book may be reprinted, reproduced, transmitted, or utilized in any form by an electronic, mechanical, or other means, now known or hereafter invented, including photocopying, microfilming, and recording, or in any information storage or retrieval system, without written permission from the publishers.

Trademark Notice: Product or corporate names may be trademarks or registered trademarks and are used only for identification an explanation without intent to infringe.

Library of Congress Cataloging-in-Publication Data

McShane-Vaughn, Mary

    Lean six sigma green belt slide book/Mary McShane-Vaughn

        p.cm.

1. Quality Control – Statistical methods – Handbooks, manuals, etc.

    I. McShane-Vaughn, Mary II. Title

    First printing 2024

Introduction

Congratulations on starting your Lean Six Sigma journey! This course manual includes a course outline, all the slides from the module videos, and the case study used in the Lean Six Sigma Green Belt course from University Training Partners. You will refer to this text throughout the course as you complete assignments and take the module quizzes. Feel free to annotate and tab the text as you go. The course manual will also prove to great takeaway and reference as you start on your first Lean Six Sigma improvement project. Enjoy!

To learn more about other Six Sigma and statistics courses University Training Partners has to offer, please visit our website: https://courses.6sigma.university

Regards,

Dr. MMV

Mary McShane Vaughn, Ph.D.
Founder
University Training Partners

Welcome!

Congratulations on starting on your Lean Six Sigma Green Belt!

A Green Belt is an important, and marketable, credential to have. Companies are looking for employees who can select the right improvement projects, manage a project team, make data driven decisions, and implement real improvements in processes. Improvements that translate into better bottom line results (i.e. $$$). As a Lean Six Sigma Green Belt, you will be able to do all of that.

This course will change the way you approach projects, the way you make decisions, and the way you see your role in your organization. Along the way, you'll have some "aha!" moments as you:

1. Get very familiar with the Six Sigma project cycle

Six Sigma concentrates on improving processes by initiating projects that follow five distinct steps: Define, Measure, Improve, Analyze and Control. These steps will roll off your tongue by the end of the course, trust me.

2. Change the way decisions are made in your organization

The decisions Green Belts make with their teams are driven by data, not by whoever has the loudest voice in the room, or the highest seniority, or the most intimidating personality. Think for a minute about your workplace: how are decisions usually made? What are the driving forces behind them?

3. Learn the right way to measure project outcomes

The goal of Six Sigma is to decrease variation – to be more consistent with our process inputs and outputs – and to decrease defects. Once we do this, we also increase customer satisfaction and improve bottom line results.

Are you excited to begin, or are you a bit scared of the math?

Sure, there is some math involved with this course. We're making decisions based on data, so there has to be some analysis, right? If you are comfortable with creating graphs and using formulas in Excel, you're way ahead of the game. If you are not, you'll be adding these tools to your already overflowing Green Belt toolbox. There are quick tutorials included in the course to help you along the way.

The truth is the hardest part of leading Six Sigma teams is not the data collection or statistical analysis – it's dealing with people. The "soft skills" of project selection, project prioritization, project scoping, stakeholder analysis, team building, and the voice of the customer are all employee-facing or customer-facing – and difficult. As I tell students, you can always hire a contract statistician to do the heavy lifting for you if you run into a complex analysis situation, but you can't outsource managing your people.

Feel better?

If you get stuck, you can always send me an email or post a question on the discussion board.

You've got this.

Now let's get started!

Course Design

Modules

This online course consists of nine separate modules that introduce you to Lean and Six Sigma and then systematically fill your toolbox with the most useful techniques for identifying and successfully running your first improvement project. You will have access to the information in these modules for 9 weeks.

Case study

Throughout the course, we'll be using a case study based on a resort hotel. This will allow you to practice applying the tools you learn in each module to a real-world situation.

Module design

For each module, there is an introduction that lets you know what you'll be learning. Then, there are a series of short videos that introduce the module material. There will be an assignment to submit, as well as a knowledge check, or quiz, for each module.

Assignments

There are short assignments for each module that will let you practice what you have learned. Many assignments will be based in Excel. Your instructor will review your submission and either accept it or turn it back to you for edits.

Don't feel bad if an assignment gets turned back to you. As Dr. MMV says, we learn the most from making mistakes. Consider a returned assignment an opportunity to grow and master a skill. Just make changes to the submission based on the instructor's detailed feedback and resubmit for another review.

Assignments are graded as 0, 1 or 2. To successfully complete the course, you'll need to score at least 26/28 total points on the 15 course assignments. .

Knowledge checks

There is also a knowledge check for each module consisting of multiple-choice questions. You can make as many attempts as needed to score at least an 80% on each knowledge check.

Instructor support

If you get stuck, or have a question or insight to share, there is a discussion board for each module. Dr. MMV monitors these boards and will answer any questions you may have. She'll also address quetions in the live weekly Zoom meetings.

Course outcomes

When you have successfully completed the nine modules, you will receive a certificate and a digital badge.

What's In It For Me?

Personal Reflection

- Take a few moments to think about what you would like to gain from this class (WiiFM).

- Write your thoughts down here, and keep this page in your notebook (we won't share these).

- We will return to your WiiFM page at the end of the course.

Planning Your Time

This instructor-led course is conducted over nine weeks. It's best to put yourself on a schedule to assure that you maintain your momentum. We've designed a weekly study plan for you to follow. Note that the last week of the course is a "grace" or catch-up week. Even if you get off track by a week or two (life does happen), you'll still have time to finish and earn that Green Belt credential!

Week	Module		Learning Activities
1	Course Introduction		Overview of the program
			Orientation to the modules
			Required textbook
			About your instructor
			FAQ
			Course calendar

Week	Module		Learning Activities
1	Module 1: Define Phase	📄	Week 1 – Define phase
		▶	Lean Six Sigma introduction
		▶	The DMAIC Roadmap
		▶	Project Selection
		▶	Voice of the Customer
		▶	Kano Analysis
		▶	Using Kano Results
		▶	The Project Charter
		⬇	Green Belt Companion spreadsheet
		✏	Discussion: Introduction
		✏	Assignment: Prioritization matrix
		✏	Assignment: Kano analysis
		🧠	Quiz 1

Week	Module		Learning Activities
2	Module 2: Define and Measure Phases		Week 2 – Define and measure phases
			Stakeholder analysis
			Using the stakeholder analysis template
			Six Sigma teams
			Process maps
			Cost of Quality
			Assignment: Stakeholder analysis
			Quiz 2

Week	Module		Learning Activities
3	Module 3: Analyze Phase	📄	Week 3 - Analyze phase
		▶	Root cause analysis
		▶	Graphical analysis
		▶	More graphical analysis
		▶	Pareto charts in Excel
		▶	Run charts in Excel
		▶	Scatter plots in Excel
		✏️	Assignment: Pareto and run chart
		✏️	Assignment: Scatter plots
		🧠	Quiz 3

Week	Module		Learning Activities
4	Module 4: Analyze Phase	📄	Week 4 – Analyze phase
		▷	Making sense of data
		▷	Central tendency and dispersion
		▷	Formulas in Excel
		▷	Normal distribution
		▷	Confidence intervals
		✏️	Assignment: Descriptive statistics
		✏️	Assignment: Z score
		✏️	Assignment: Confidence interval
		🧠	Quiz 4

Week	Module		Learning Activities
5	Module 5: Analyze Phase	📄	Week 5 – Analyze phase
		▶	Process capability
		▶	Process capability problems
		✏️	Assignment: Process capability
		🧠	Quiz 5

Week	Module		Learning Activities
6	Module 6: Improve and control phases	📄	Week 6 – Improve and control phases
		▷	Lean tools
		▷	Eight wastes/ quick changeover
		▷	Formula 1 pit stops
		▷	Meals per hour
		▷	Value stream mapping
		▷	Kaizen event planning
		✏️	Discussion: Lean implementation
		✏️	Assignment: Mistake proofing
		✏️	Assignment: Eight wastes
		⚙️	Quiz 6

Week	Module		Learning Activities
7	Module 7: Control phase		Week 7 – Control phase
			Control plans
			Control charts
			Which SPC chart
			Assignment: Control charts
			Quiz 7

Week	Module		Learning Activities
8	Module 8: Putting the pieces together		Week 8 – Putting the pieces together
			FMEA
		▷	Putting the pieces together
		▷	Project reports – Jim Sutton
		▷	Project reports – Eli Guzman
		⬇	Project summary template
		⬇	FMEA article
		✏	Discussion: What's in it for me reflection
		⚙	Quiz 8

MIRASOL RESORT

A Lean Six Sigma Case Study

© 2024 University Training Partners

JONATHAN SAND

President Sand Island Resorts

This case is based on a small beach resort hotel recently acquired by Jonathan Sand, President of Sand Island Resorts, LLP. Sand began working in the hotel industry as a teenager, manning the front desk and taking reservations at the local hotel owned by his uncle.

After earning his college degree in manufacturing management, he worked as an assembly line supervisor at an automotive plant as well as a production engineer in the pressed metal shop. When the assembly plant was shut down, Jonathan returned to the East Coast and found a position as a maintenance supervisor at Mirasol, a local beach resort. He found that his industrial background gave him a unique edge in attacking and solving problems in the hospitality sector.

In the past five years, he has held wide-ranging positions at the resort to include accounting supervisor, banquet manager, and housekeeping manager. In 2018, Sand was promoted to general manager of Mirasol, which was then owned by Allendale Holdings, a company that managed seven high-end resorts in South Carolina and Florida.

VISIT SOUTH CAROLINA

Takeover from Allendale Holdings

The effect of the pandemic on luxury travel forced Allendale to liquidate some of its properties at deeply discounted prices in order to remain solvent. When it was announced that Mirasol was being sold off, Jonathan contacted his uncle, now retired, and formed the Sand Island Resorts partnership. Sand and his Uncle Fred were able to acquire the property a year ago. Since then, Sand has been enthusiastically making changes to put his own stamp on the resort.

In 2022, the resort was featured on the Condé Nast Traveler newletter, which described it as "a boutique resort that is the ultimate in a relaxing getaway; private and personalized."

> "A BOUTIQUE RESORT THAT IS THE ULTIMATE IN A RELAXING GETAWAY; PRIVATE AND PERSONALIZED."

Inspired by the Condé Nast review, Sand crafted the new company mission statement and has it displayed in framed posters that hang in the main lobby and the back offices:

"Sand Island Resorts is committed to providing its member guests with personalized service and a relaxing resort experience that consistently delivers excellent quality and high value."

FACILITIES

The Mirasol resort is located on a small barrier island along the Intercostal Waterway, accessible only by passenger ferry. The ferry, operated by the State, makes six round trips each day. Automobiles are restricted on the island, and the main means of transportation are golf carts and bicycles.

The small resort has 50 suite-style rooms in the two-story main building. There are two pools, one with a swim-up bar, as well as three heated spas on the property. Water sports such as jet ski and sail boat rentals are accessible from the marina located to the west of the property, reachable by a complimentary golf cart shuttle. The beach is accessible via a boardwalk, and guests can reserve cabanas, umbrellas and chairs, surf boards, boogie boards, and snorkel equipment at the beach pavilion.

There is one restaurant serving breakfast and dinner daily in the main building, a small bar located off of the main lobby, a poolside snack bar, and a separate banquet/meeting facility that can accommodate up to 175 guests. A gift shop is in the lobby, selling sundry items and resort-branded tote bags, beach towels, golf shirts and Tervis tumblers. A separate exercise and spa facility is located next to the pool area. The gym has four stationary bikes, three treadmills, a Stairmaster, an aerobics/yoga room and locker rooms with showers. The spa has two treatment rooms, three manicure and pedicure stations and two outside massage cabanas.

The facilities building is located at the north end of the property, and contains a commercial laundry facility, a maintenance garage for the resort golf carts and bicycles, a tool shop, two storage rooms, a hurricane preparedness room, and the engineering department offices.

ECONOMIC & MARKET PRESSURES

Allendale Holdings saw a 35% decline in room occupancy and special events bookings at the start of the pandemic in 2020. Since purchasing the resort, Sand Island Resorts has experienced a 5-7% growth in these rates.

Sand attributes the growth to targeted online ads he ran on social media in the Northeast, and would like to expand the advertising campaign. His uncle would rather concentrate advertising efforts on adjoining states due to the cost of gasoline.

Reina Andersen, Marketing Director, has proposed partnering with influencers specializing in luxury travel. Sand is reluctant to move forward due to recent scandals involving several influencers.

In a recent meeting, the Director of Recreation and Spa Services, Linda Petrucelli, suggested offering a spa destination package during the "shoulder season," complete with spa cuisine, spa treatments, and weeklong courses taught by celebrity yoga masters. Sand thinks that demand for such event is not strong enough to offset the cost.

While Allendale was experiencing a cash flow shortage, it postponed routine maintenance and upkeep. Upon acquiring the property, Sand had to perform costly repairs to the pool deck, air conditioning system and septic system.

In addition, Sand Island Resorts repainted the exterior, made improvements in the restaurant and spa facilities, and purchased new furniture for the beach cottages.

Over the past year, the cost of maintenance, repairs and improvements have eaten into SIR profits. Controller Greg Schultz has suggested reducing the number of resort amenities to offset increased expenses, but Sand is worried that fewer services will translate into fewer bookings.

In addition to economic worries, Sand is concerned that a new hotel recently opened on the south end of the island will dilute his market share. The hotel is an all-inclusive resort owned by a national chain that caters to families with children. He has a suspicion that that his more price sensitive customers are flocking to the new hotel because it offers more activities and lower prices. In addition, couples seeking a destination wedding venue may be enticed by the larger banquet facility. This would be a huge blow to Mirasol's revenue, since weddings make up 20% of its bookings from April through October.

Appointment bookings at the spa have dropped, making it difficult for Sand to justify keeping his three aestheticians on staff. In previous years, the spa was a profit center for the resort. Sand feels that the spa is disorganized and suffers from a lack of consistency in atmosphere: depending who is at the desk, the customer experience could range from muted and distant to boisterously friendly.

Reina Andersen has been pushing to expand the merchandising business by marketing a line of Mirasol stemware and bar accessories to sell through department stores such as Dilliard's and Nordstrom. She feels this will create new interest for the target market and will be a natural addition to the wedding gift registries of couples married at Mirasol. Both Sand and Greg Shultz, the Controller, see potential in the idea, but are reluctant to make the capital investment for the new line given the current economic climate.

OPERATIONAL ISSUES

According to Eli Guzmán, the Housekeeping Manager, water usage in the on-premises laundry facility has increased by 25% compared to last year. Over the past few months, he has worked with the service contractor to adjust the wash cycle parameters, but the changes have not reduced water usage to target levels. Guzmán has suggested switching to a contractor that leases wash water recycling systems, but Greg Schultz has rejected the idea based on the high installation fee and monthly leasing costs.

The golf cart maintenance garage has experienced parts shortages in the past six months due supplier problems in Japan. This past summer, three of the ten large, 8-passenger golf carts were out of commission, which caused delays in shuttling guests to and from the ferry dock. Chief Engineer Jim Sutton would like to stockpile replacement parts to avoid shortages in the future, but the garage space is already too small for the number of vehicles they service, and parts and tools are often difficult to find.

Inclement weather or equipment problems can cause delays in the ferry schedule, which can leave Mirasol understaffed at crucial times. Sand has adopted an "all hands on deck" approach when these incidents occur, but knows that the quality and speed of the work performed suffers. Last Friday, he and his uncle cleaned 4 rooms when the ferry was two hours late due to engine trouble. Later, his housekeeping staff told him that he had missed vacuuming under the beds, failed to replenish the Tazo tea bags, and did not fold the duvet according to standard.

There has been high absenteeism and significant turnover in the maintenance and housekeeping departments in the past 6 months. Sand believes that the new resort may be offering higher pay. Training new staff members has taken time away from other duties of the Chief Engineer and Housekeeping Manager.

After the sale of Mirasol, the Executive Chef and Banquet Manger decided to remain with Allendale Holdings and were reassigned to a

PAGE 07

resort on the Gulf Coast of Florida. Since then, Frank Soule, the sous chef, has taken over the Executive Chef and Banquet Manager duties while an outside firm conducts a talent search. Frank has handled his new responsibilities reasonably well, but Sand fears that he may quit due to exhaustion if new staff members are not found quickly.

The quality of service has suffered recently. In the past, an average of 4 complaint cards per month was received by the front desk. In June, 25 complaints were filed by guests. Customers have complained about lengthy waits for the shuttle, special requests that were ignored, incorrect room service orders, inoperable air conditioners and overcharges upon check out. The overcharges have particularly grated on Sand, who wants to make sure this mistake never happens again.

SIX SIGMA IMPLEMENTATION

Sand was involved in quality initiatives at the automotive plant where he previously worked, and had met with the plant's quality manager to discuss data collection and SPC charts a few times. He had tried to incorporate some quality principles into the business operations while he was general manager of the resort back in the Allendale days, but he had neither the experience nor the corporate support to implement a full-fledged program. While attending a hotel and hospitality convention last fall, Sand heard a presentation from an owner of a mid-sized hotel that had decreased costs and increased occupancy by implementing Six Sigma and Lean techniques. Sand began researching Six Sigma and decided to use the DMAIC process to address some of his resort's problems. He himself became Green Belt certified, along with the housekeeping manager and chief engineer.

The other directors and managers have gone through a 2-day Yellow Belt course, led by an outside consultant who is a Six Sigma Black Belt. With the Black Belt's guidance, SIR plans to identify and complete 3-4 Six Sigma projects this year.

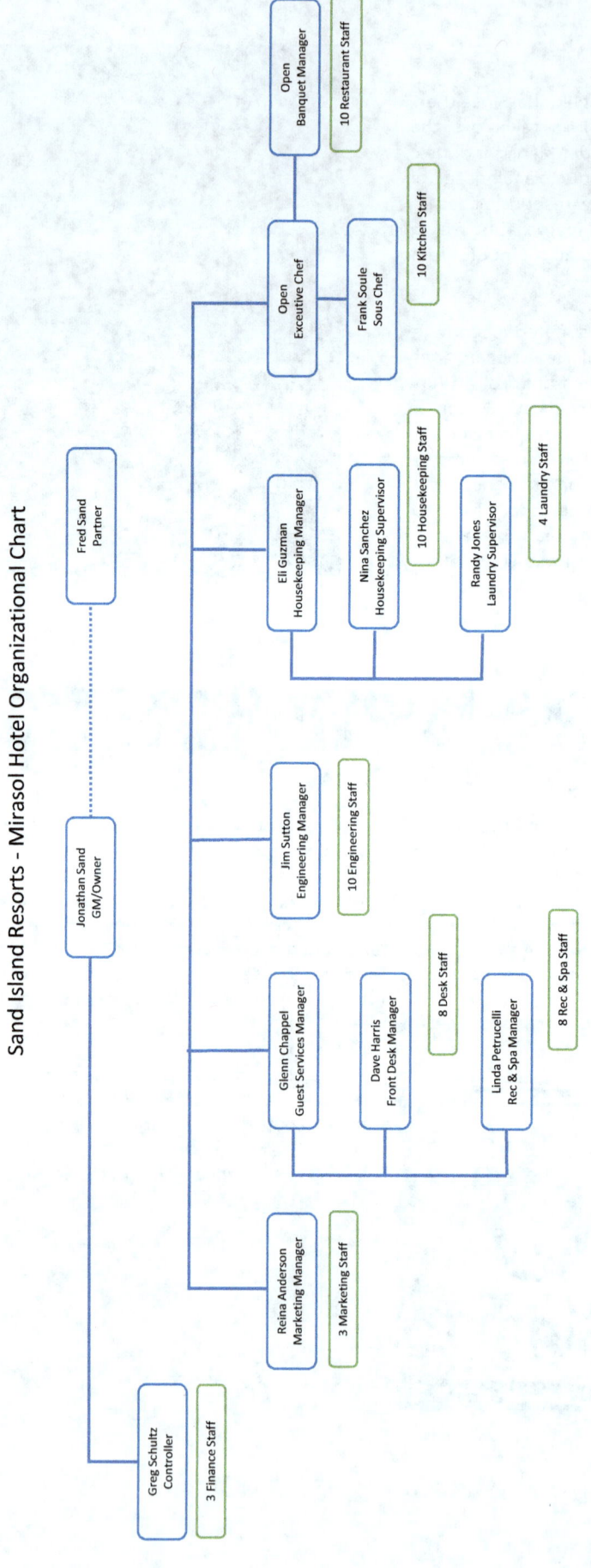

Sand Island Resorts - Mirasol Hotel Organizational Chart

Data Tables prepared by Greg Schultz, Reina Anderson and Dave Harris

All data from previous year

Occupancy rate by month

January	10%
February	15%
March	20%
April	45%
May	50%
June	75%
July	70%
August	65%
September	65%
October	45%
November	20%
December	30%

Average Length of Stay in days by month

January	1.6
February	1.7
March	2.8
April	3.6
May	2.5
June	3.5
July	3.9
August	3.6
September	3.7
October	1.7
November	1.5
December	2.6

Revenue per Room by month

January	$ 1,395
February	$ 1,890
March	$ 3,255
April	$ 7,628
May	$ 9,300
June	$ 17,438
July	$ 17,360
August	$ 16,120
September	$ 12,188
October	$ 7,324
November	$ 2,400
December	$ 4,883

Percent Bookings by reservation channel

Hotel website	54%
Direct Calls	16%
Online booking agencies	30%

Total Room Changeovers by Month

January	78
February	99
March	89
April	150
May	248
June	257
July	223
August	224
September	211
October	328
November	160
December	143

Note: A changeover occurs when a guest checks out of a room and a new guest checks in

Total complaints by month

January	2
February	2
March	4
April	4
May	5
June	25
July	20
August	24
September	18
October	27
November	15
December	10

Percent complaints per month by room changeover

January	3%
February	2%
March	4%
April	3%
May	2%
June	10%
July	9%
August	11%
September	9%
October	8%
November	9%
December	7%

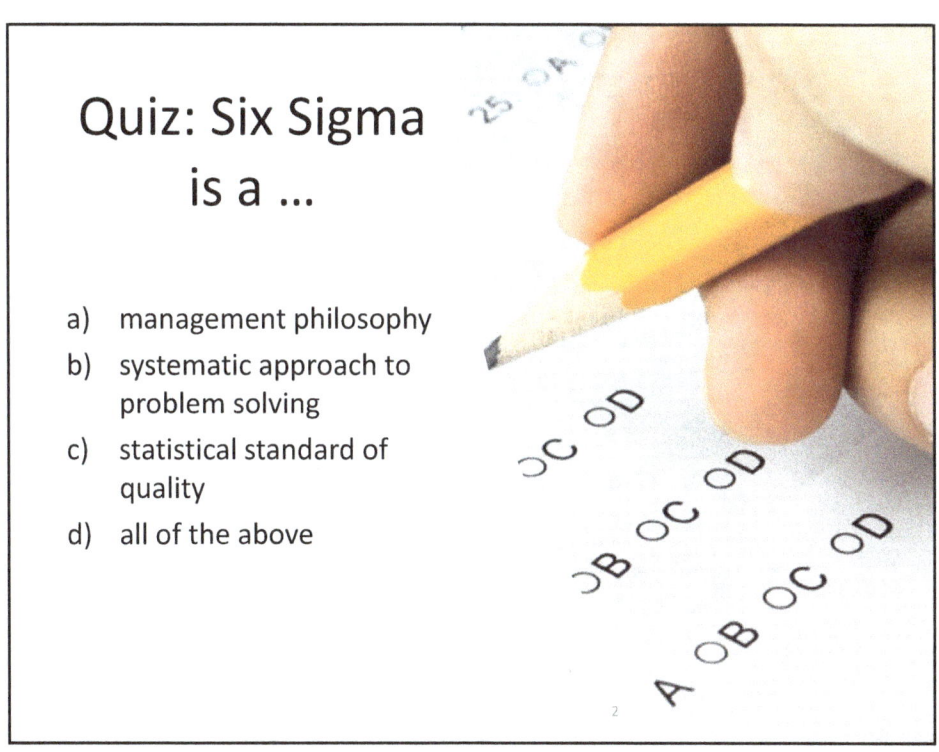

a) Management Philosophy

- Organization-wide deployment and involvement
- Driven from the top down
- Elevated standard of excellence
- Linked to the business' bottom line
- Requires in-depth training of quality tools
- Oriented toward projects

b) Systematic Approach to Problem Solving

- Process focused

    A process is a sequence of steps that uses inputs and produces a product or service as an output.

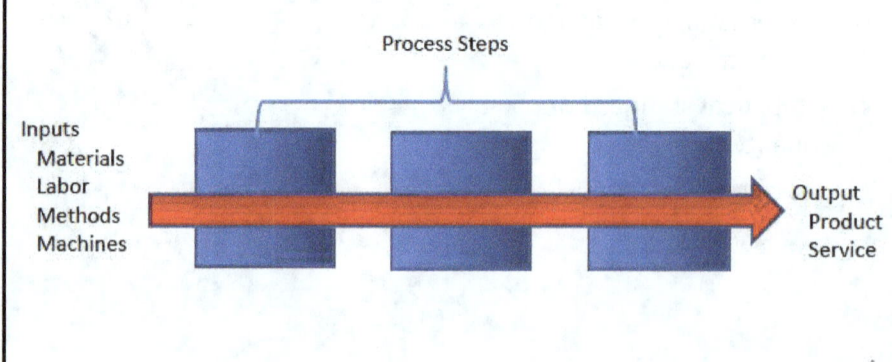

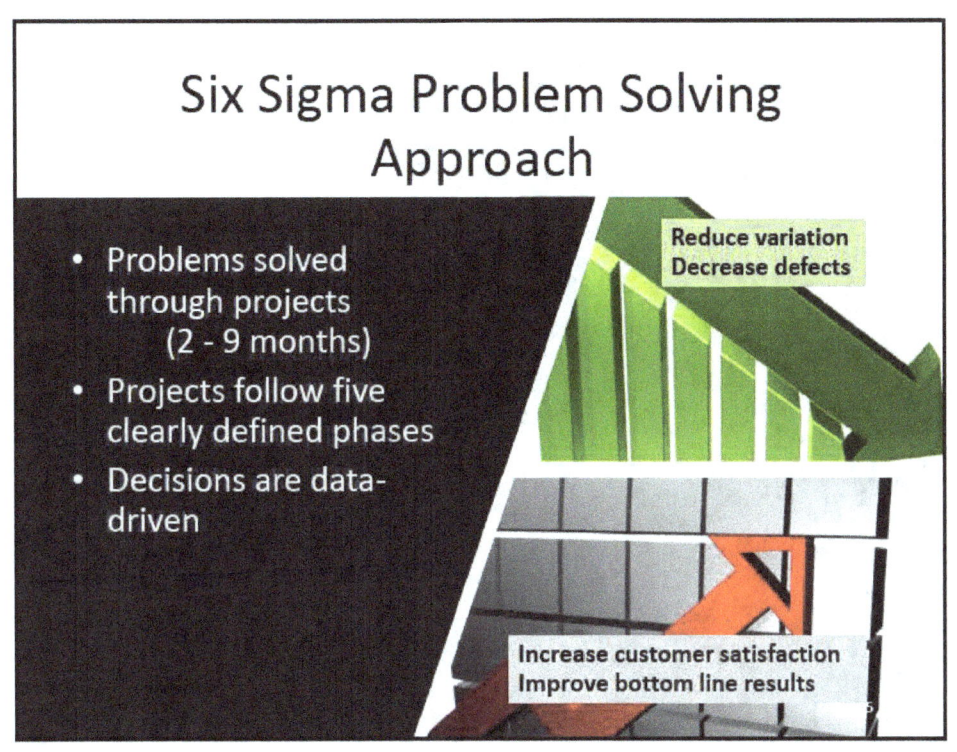

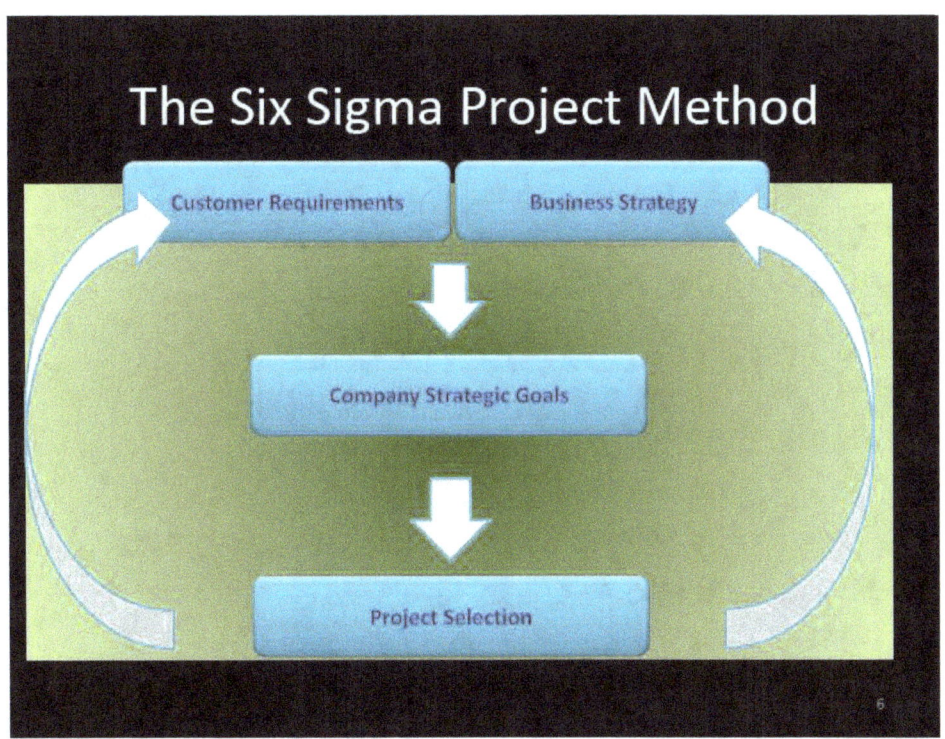

c) Statistical Standard of Quality

- A Six Sigma (6 σ) level of quality translates into

    3.4 defects per million opportunities (dpmo)

6σ

Six Sigma: When Great Isn't Good Enough

What does the 6 sigma = 3.4 dpmo standard of quality really mean?

In April 2010, a major hamburger chain served approximately **1.26 billion** hamburgers.
- At a 99% quality level = 3.8 sigma
    12,600,000 undercooked hamburgers
- At a 99.99966% quality level = 6 sigma
    4,284 undercooked hamburgers

More Examples

99% Quality = 3.8 sigma	99.99966% Quality = 6 sigma
• 7,011 vehicle breakdowns per day	• 870 breakdowns per year
• 2,441 boating distress calls per month	• 10 distress calls per year
• 3,540 dropped babies per week	• 5 dropped babies per month
• 1,452 surgical mistakes per day	• 180 surgical mistakes per year

With 1.5 sigma shift

A Few Actual Sigma Levels

Actual Sigma Levels and DPMO for Selected Activities

- Flight safety — ~6.1 sigma
- Baggage handling — ~4.2 sigma
- AT&T calls — ~3.1 sigma
- IRS helpline — ~3.0 sigma (Typical US company)
- Airline on-time arrivals — ~2.3 sigma

With 1.5 sigma shift

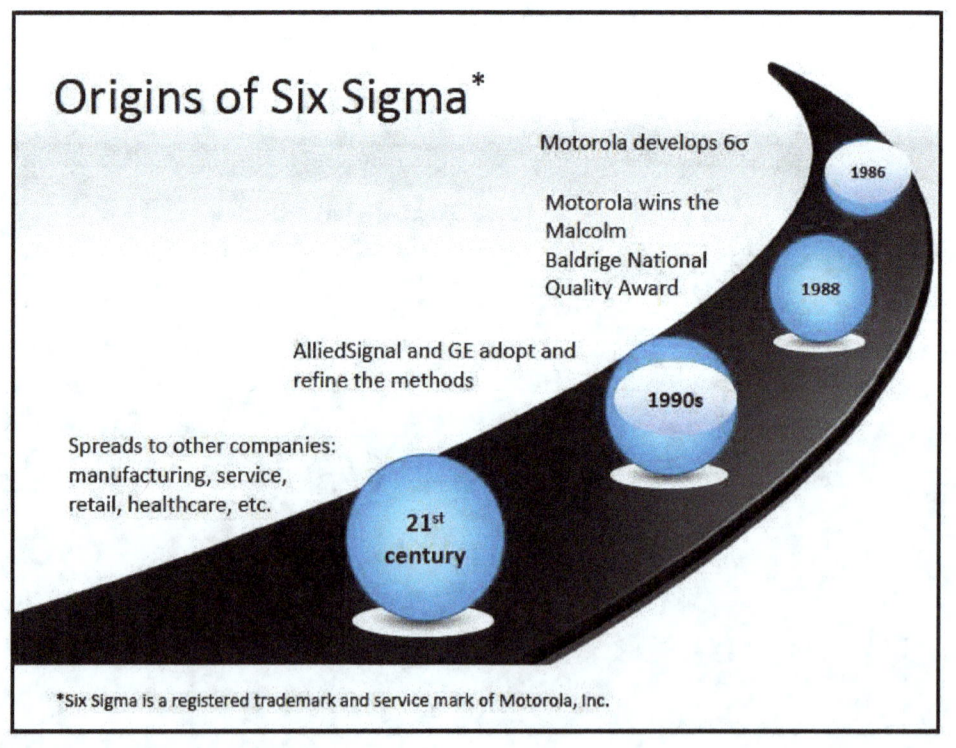

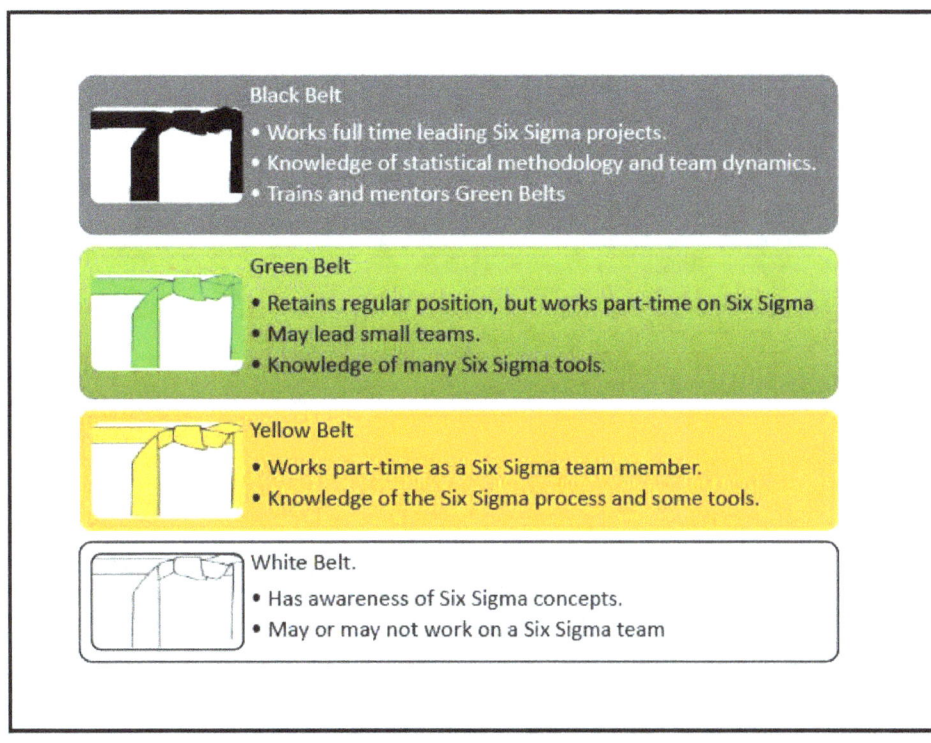

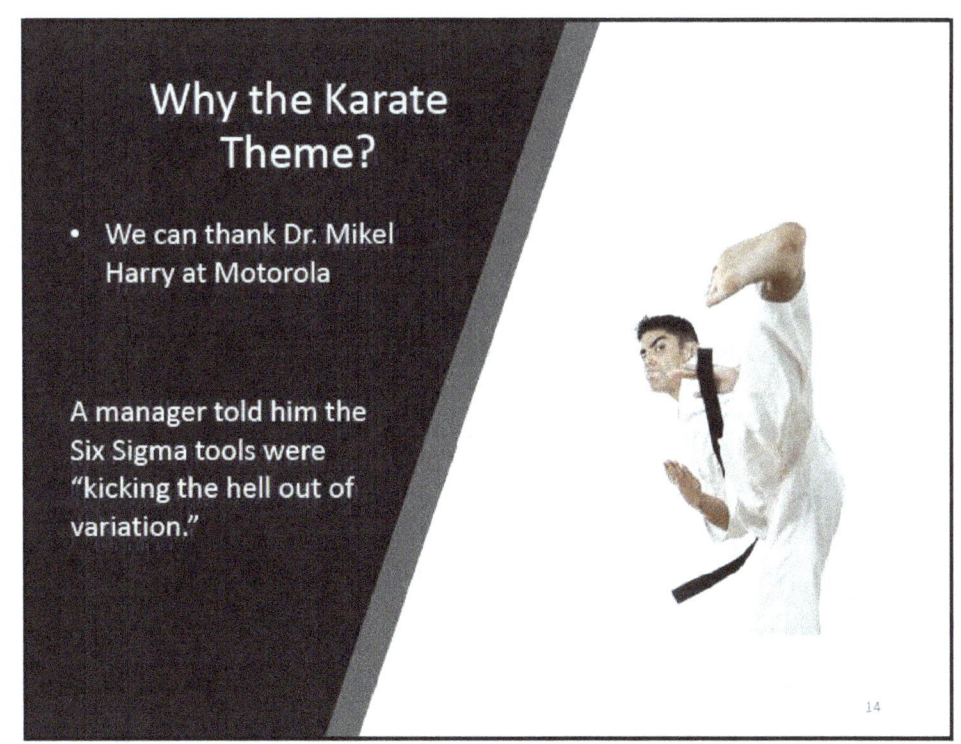

More Six Sigma Roles

- Master Black Belt
  - Full time position
  - Trains and mentors Black Belts
  - Works with management to choose Six Sigma projects
  - Advanced knowledge of statistical techniques

- Champion or Sponsor
  - Top-level manager familiar with Six Sigma
  - Supports teams by providing resources and removing barriers

- Executive
  - Shows support of Six Sigma through communication and actions
- Process Owner
  - Team member or leader
  - Manager responsible for all aspects of a process with the authority to make changes
  - Should be at least a Green Belt

Six Sigma Features

- Enterprise wide
- Top down
- Data driven
- Project oriented
- Extremely high standard for quality

- **BOTTOM LINE RESULTS**
  Something top management can get behind

Some Six Sigma Results

According to the Six Sigma Academy...

Black Belts save companies approximately $230,000 per project and can complete 4 to 6 projects per year.

General Electric, one of the most successful companies implementing Six Sigma estimated benefits of around **$10 billion** during the first five years of implementation.

Lean

- ✓ Common sense approach
- ✓ Hands on training
- ✓ Little to no math
- ✓ Quick, dramatic results

Transportation
Inventory
Motion
Waiting
Overproduction
Over-processing
Defects
Skills

- Emphasis on <u>reducing waste</u>, continuous improvement

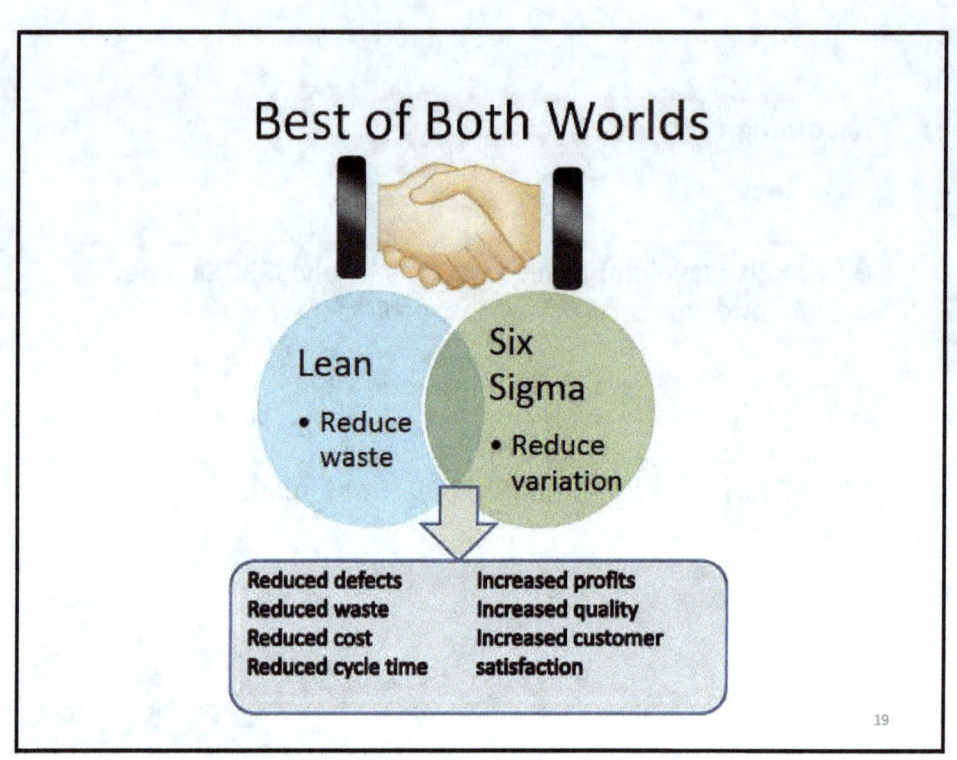

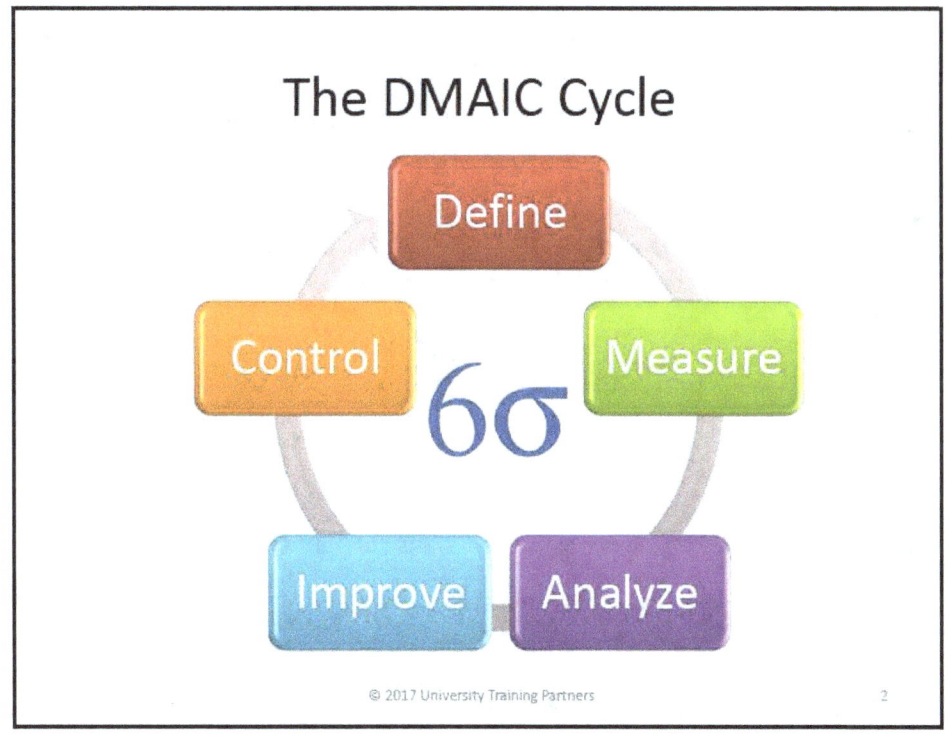

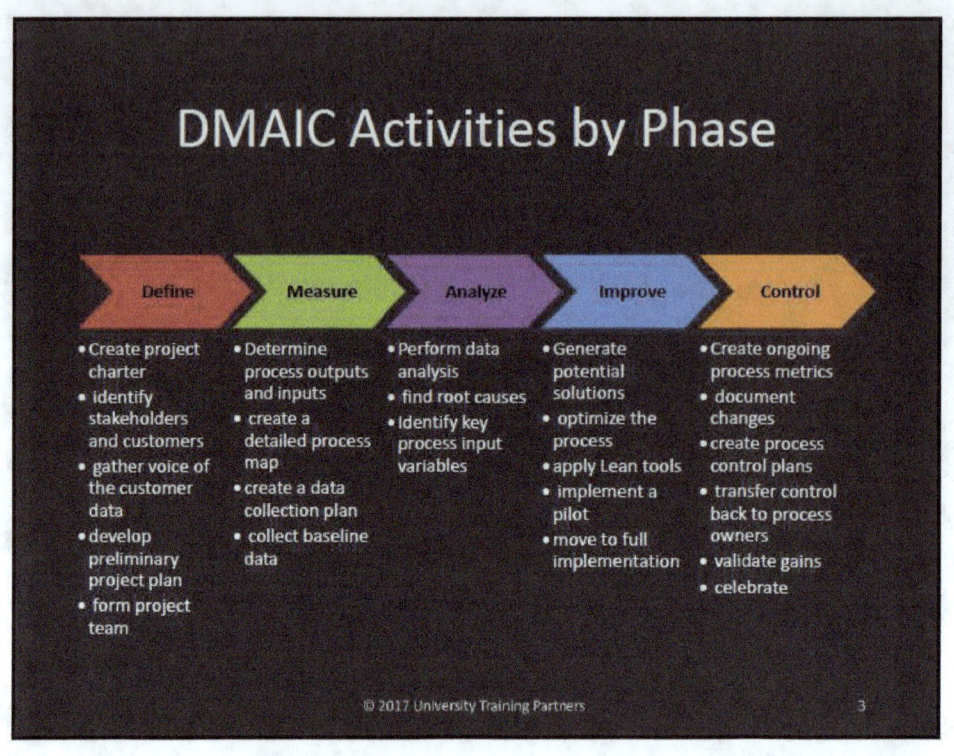

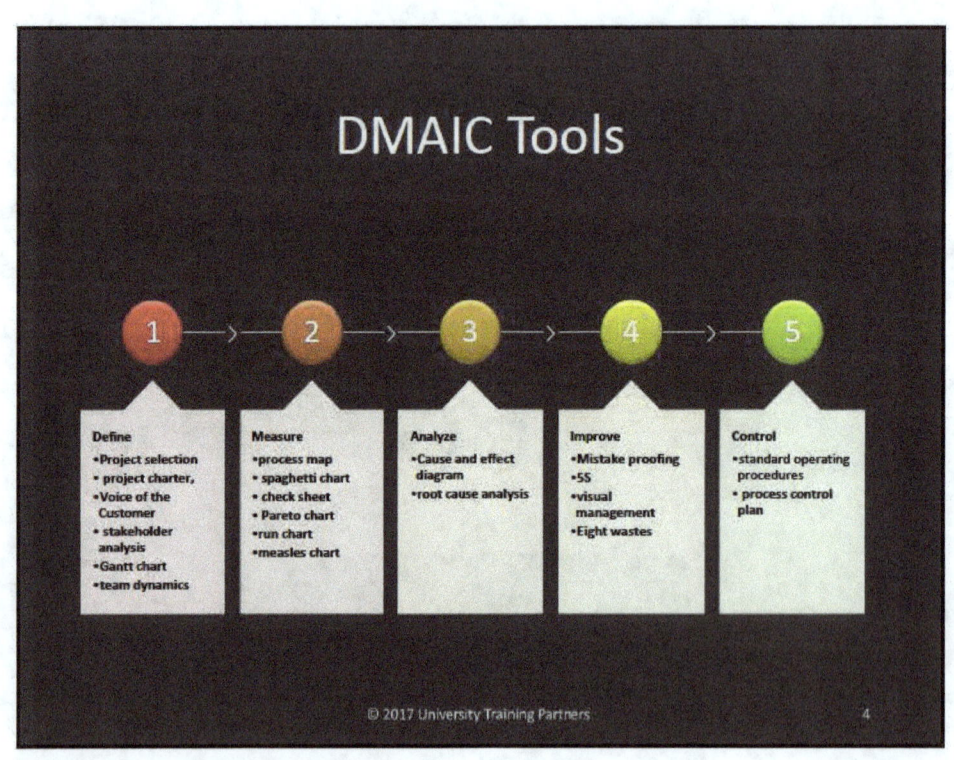

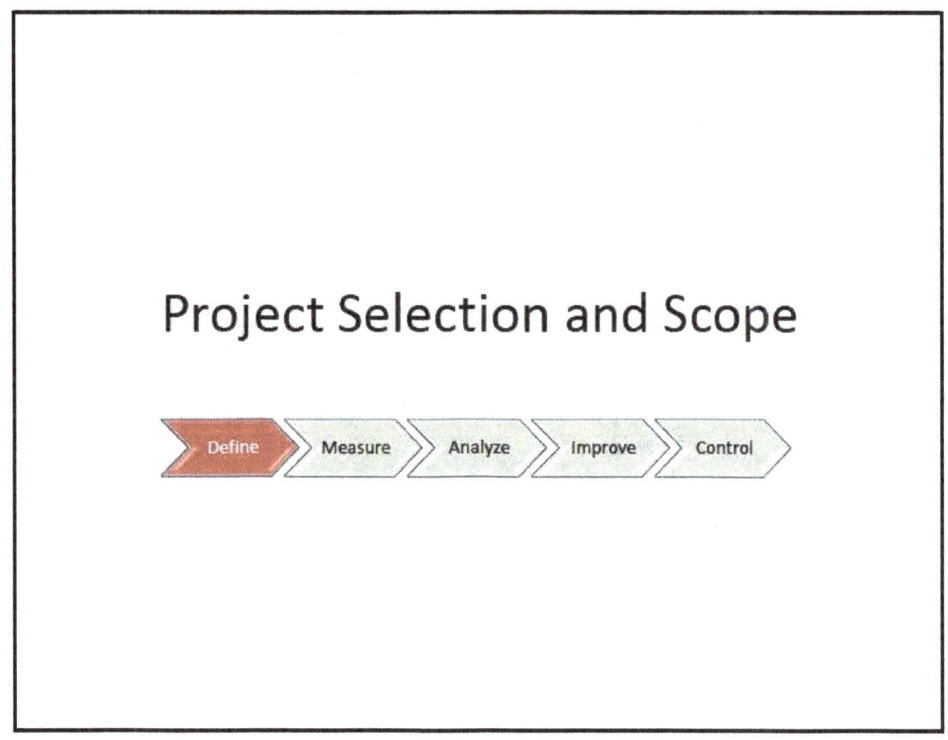

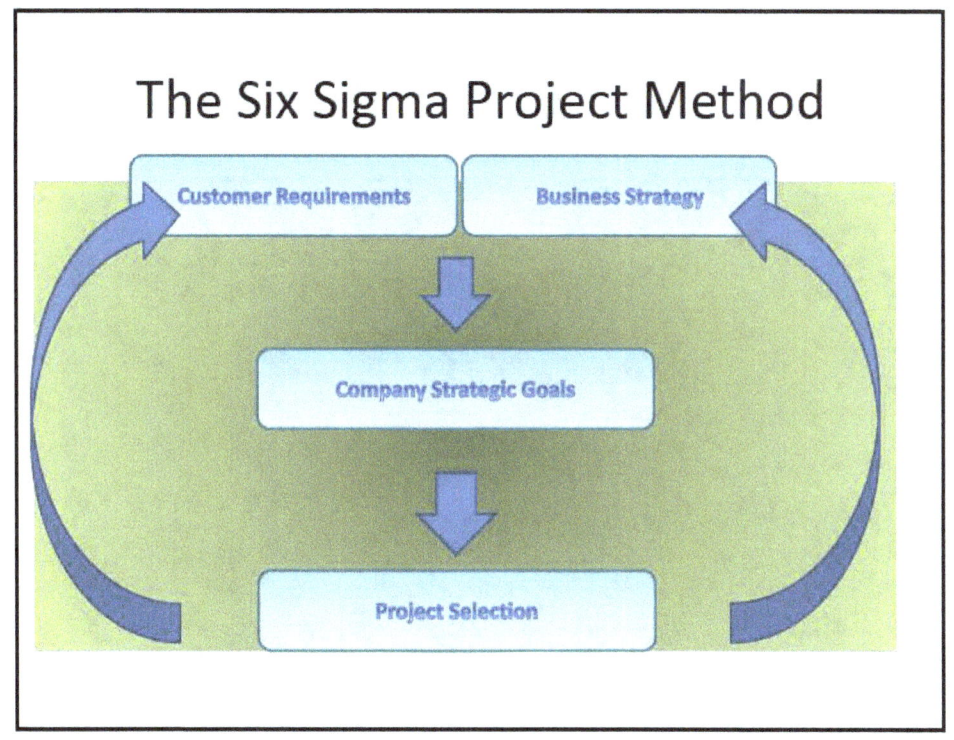

Selecting Projects

- Performed by a steering committee

- Considerations
  - Does the project tie into the company strategic plan?
  - Does the project address customer needs?
  - What is the probable duration of the project?
  - What is the expected cost-benefit ratio?
  - What is the project's level of complexity?
  - Does the company have the resources available for the project?

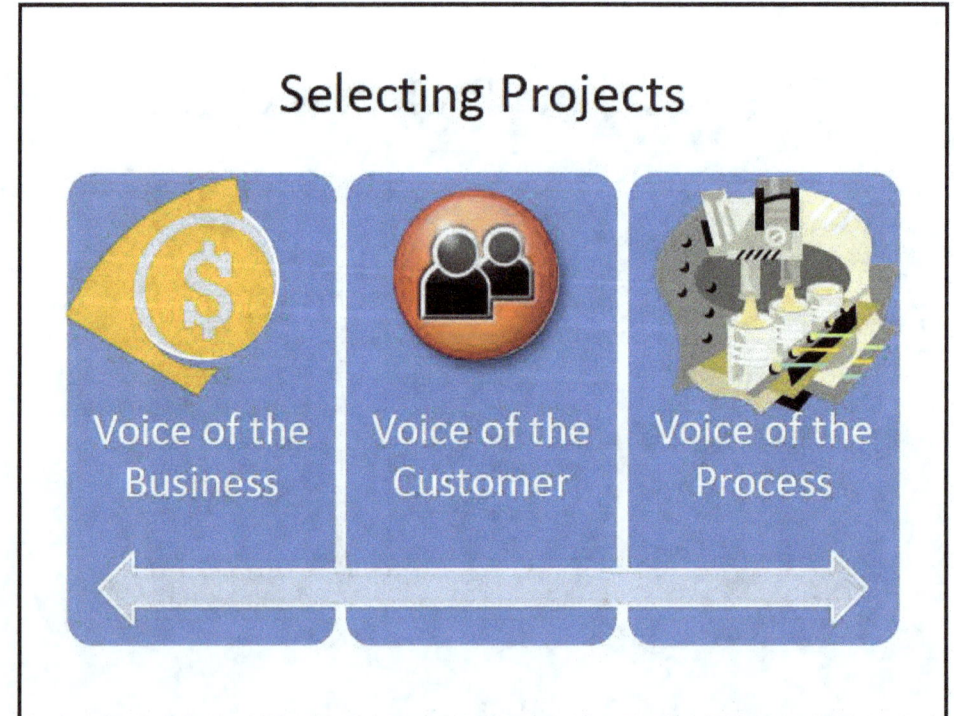

Selecting Projects

- Environmental Health and Safety
- Government, Regulatory
- Voice of the Employees

Prioritization Matrix Example

Choosing Among Office Supply Vendors

Use a **prioritization matrix** to compare various options based on set criteria

Criteria	Selection	Cost	Speed	Support
Relative Importance	25%	40%	20%	15%
Option				
Vendor A	1	1.5	3	1
Vendor B	3	3	2	2
Vendor C	2	1.5	1	3

Criteria	Selection	Cost	Speed	Support	
Relative Importance	25%	40%	20%	15%	
Option					Total
Vendor A	0.25	0.6	0.6	0.15	1.60
Vendor B	0.75	1.2	0.4	0.3	2.65
Vendor C	0.5	0.6	0.2	0.45	1.75

Project Scope

- Project should take 2-9 months to complete
- Don't try to solve world hunger...
  - "Improve customer satisfaction"
  - "Reduce warranty claims"
- Don't make scope so narrow that it prevents the root cause from being found
  - "Decrease wait times for customers by increasing the number of support staff."

Capturing the Voice of the Customer

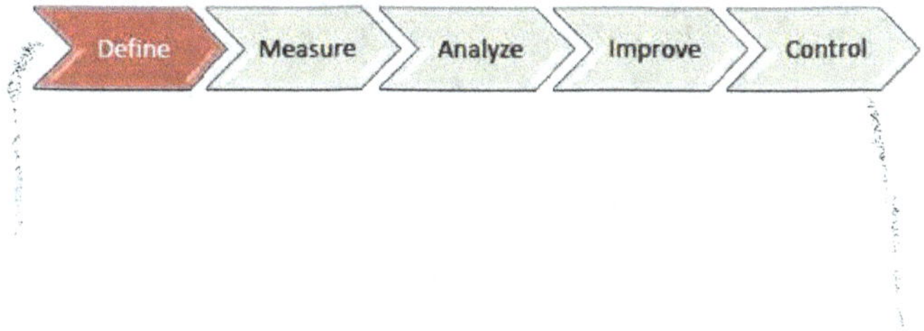

Define > Measure > Analyze > Improve > Control

Who is the Customer?

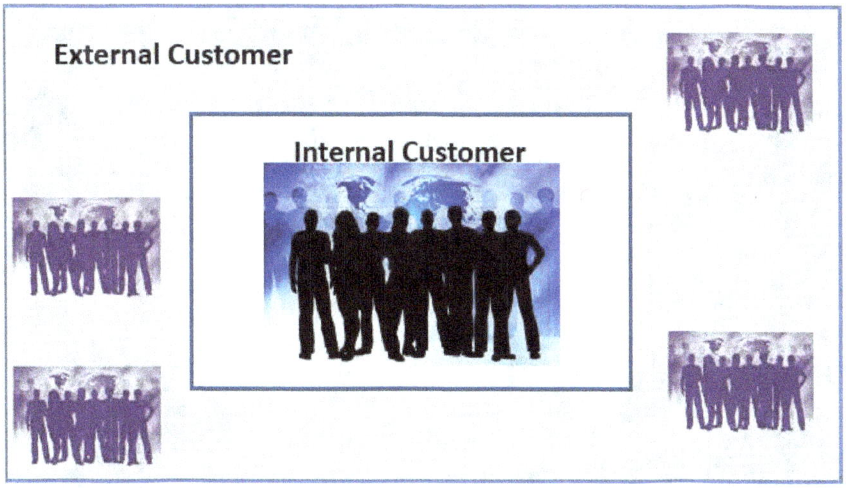

External Customer

Internal Customer

VOC

- Listen to the Voice of the Customer in order to

  – Discover what customers care about (and what they are willing to pay for)

  – Align corporate and project goals and priorities with customer needs

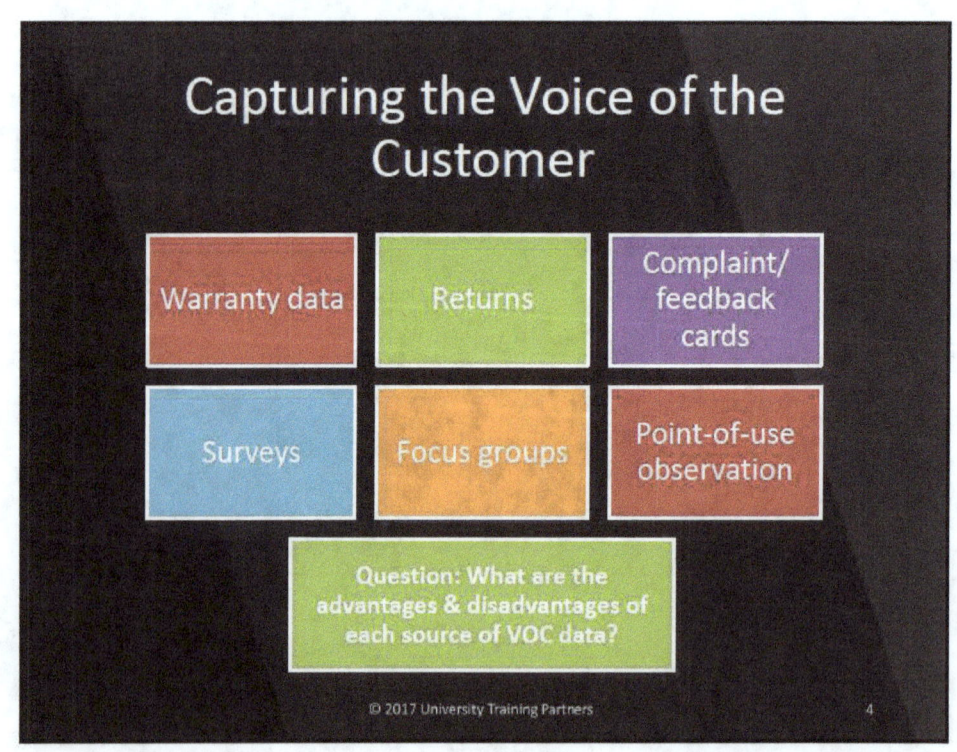

Translating the VOC into ...

CTQ = Critical-to-Quality Requirements

Dashboards

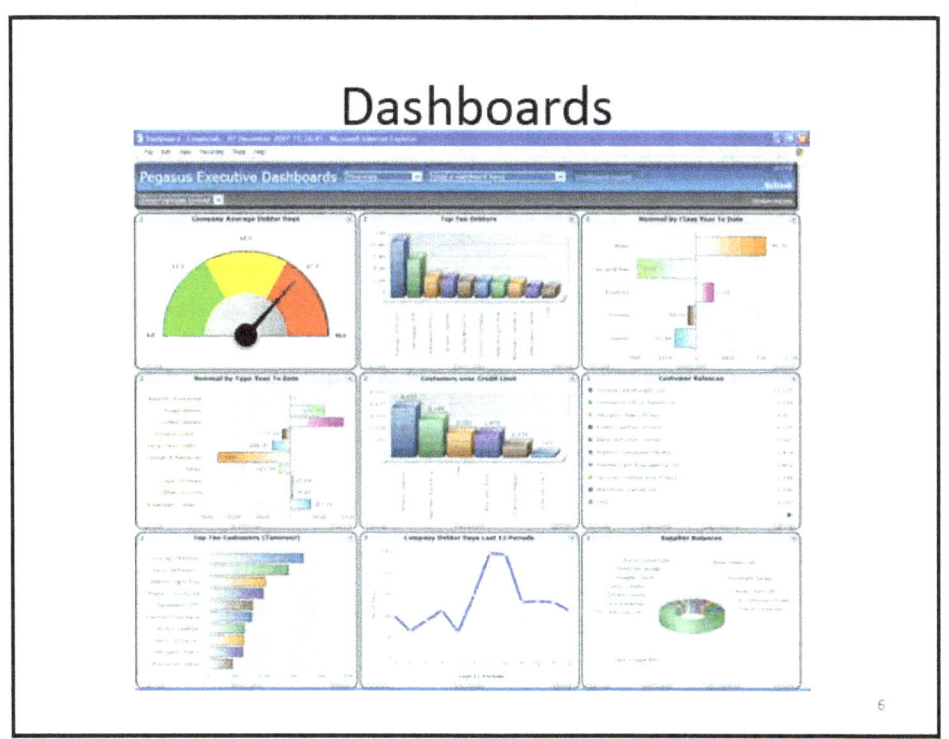

Kano Analysis

Kano Analysis

- Customers have specific needs and wants.

These can be spoken or unspoken. They can even be unknown to them.

- **Satisfiers** – performance features
- **Must-haves** – basic requirements
- **Delighters** – excitement features

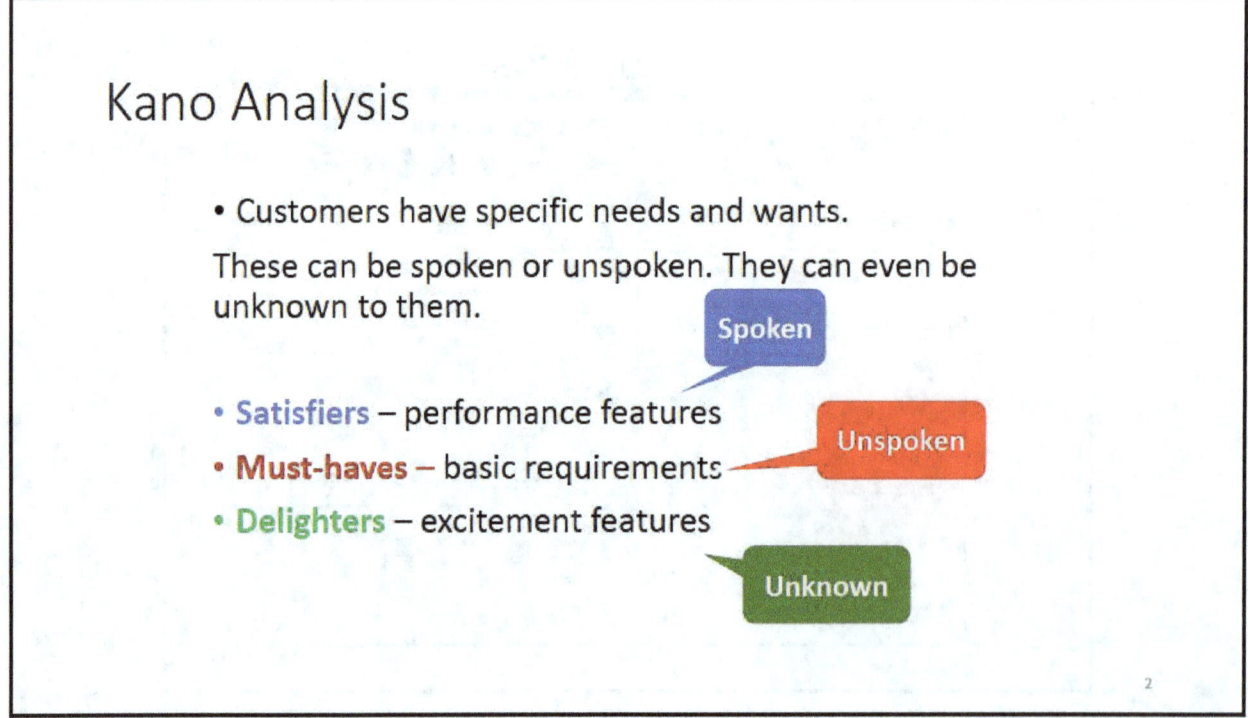

The Kano Model

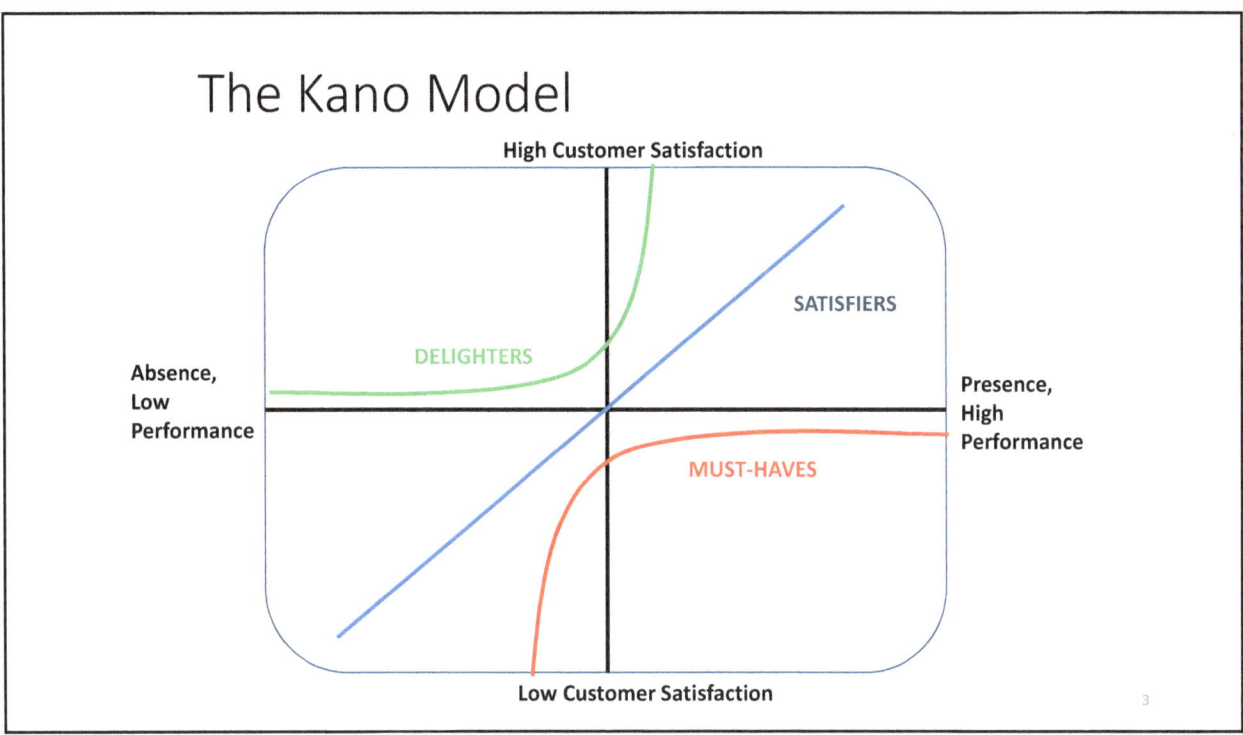

Determining Kano Classifications

Kano categories can be determined through specialized surveys using two questions for each feature.

***Presence*, or *functional question*:**
If the bed linens in a hotel are of high quality, how do you feel?
- Happy, I **like** it that way
- Neutral, I **expect** linens to be high quality
- Neutral, high quality linen is not important to me
- I can **live with** it
- Unhappy, I **dislike** it that way

***Absence* or *dysfunctional question*:**
If the bed linens in a hotel are not of high quality, how do you feel?
- Happy, I don't **like** high quality linens
- Neutral, I don't **expect** high quality linens
- Neutral, high quality linen is not important to me
- I can **live with** it
- Unhappy, I **dislike** low quality linens

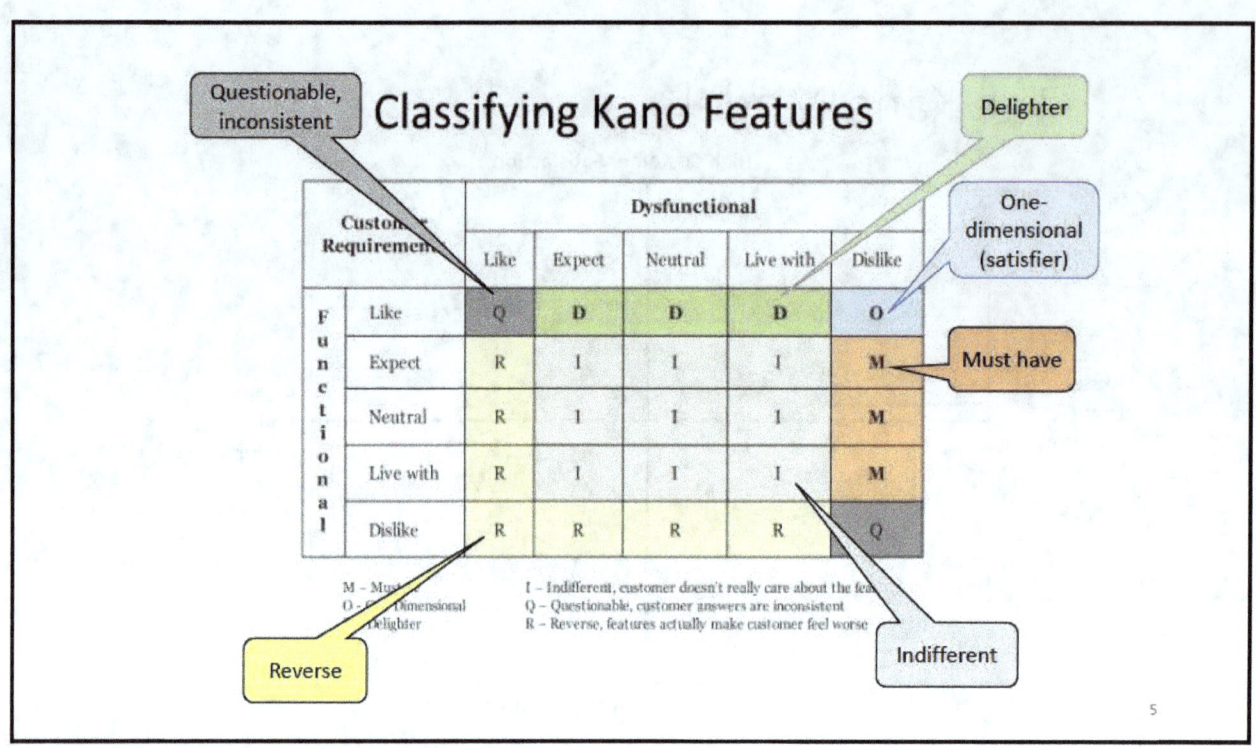

Using Kano Survey Results

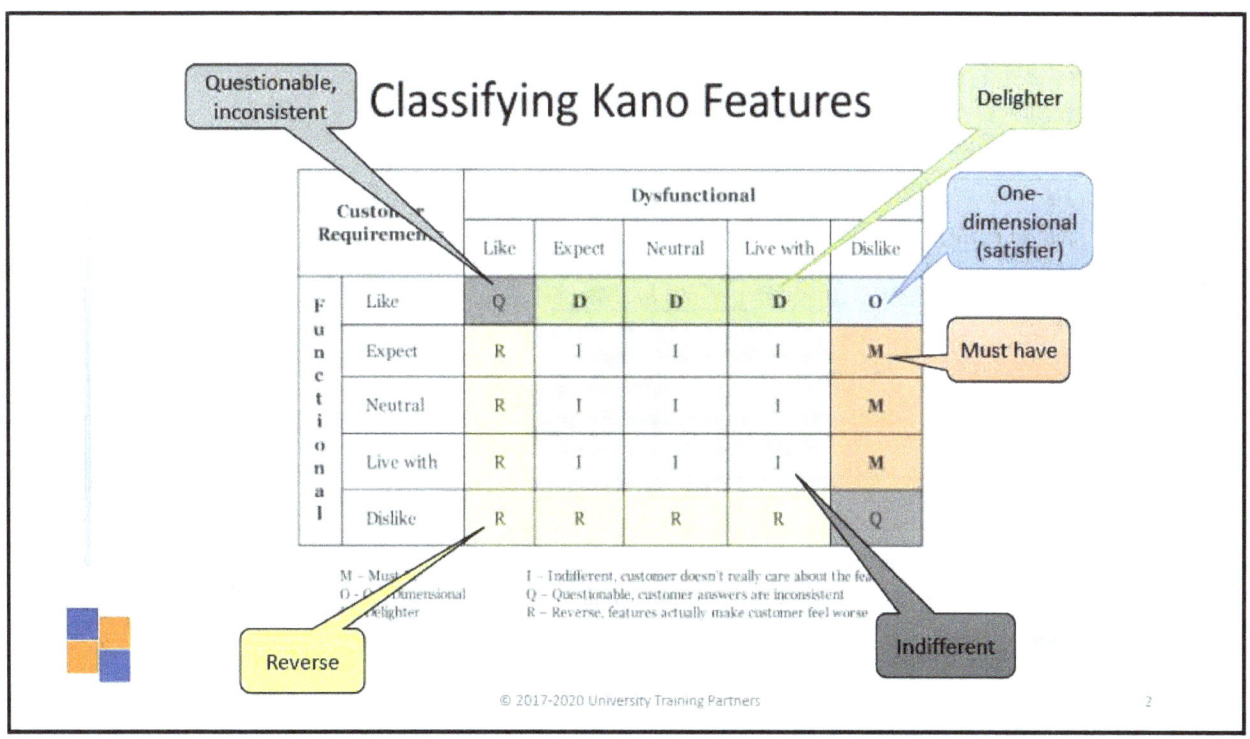

Individual Responses

Calculated cells	
INDIVIDUAL RESULTS - Your name here	
Feature	Kano Designation
High quality bed linens	S
High quality restaurant meals	S
Swim-up bar	I
Room service	I
Spa services such as mani/pedis	D
Nightly entertainment	I
Water sports	D
A high cleanliness score	M
High quality toiletries	I
Free wifi	M

Kano Survey Results

GROUP TALLY

Feature	Must Have	Satisfier	Delighter	Indifferent	Reverse	Questionable
High quality bed linens	35%	33%	12%	20%	0%	0%
High quality restaurant meals	21%	22%	25%	32%	0%	0%
Swim-up bar	0%	0%	15%	75%	10%	0%
Room service	10%	7%	0%	65%	8%	0%
Spa services such as mani/pedis	0%	11%	15%	70%	4%	0%
Nightly entertainment	11%	6%	16%	67%	0%	0%
Water sports	0%	6%	34%	60%	0%	0%
A high cleanliness score	53%	45%	0%	0%	0%	2%
High quality toiletries	19%	11%	55%	15%	0%	0%
Free wifi	67%	19%	7%	7%	0%	0%

The Project Charter

Define → Measure → Analyze → Improve → Control

Project Charter

- Outlines project purpose, benefits, scope, expected results
- Components
  - Project Title
  - Problem Statement
  - Goal Statement
  - Baseline Metrics and Goal Metrics
  - Team members and roles
  - Milestones and completion dates
  - Resources needed

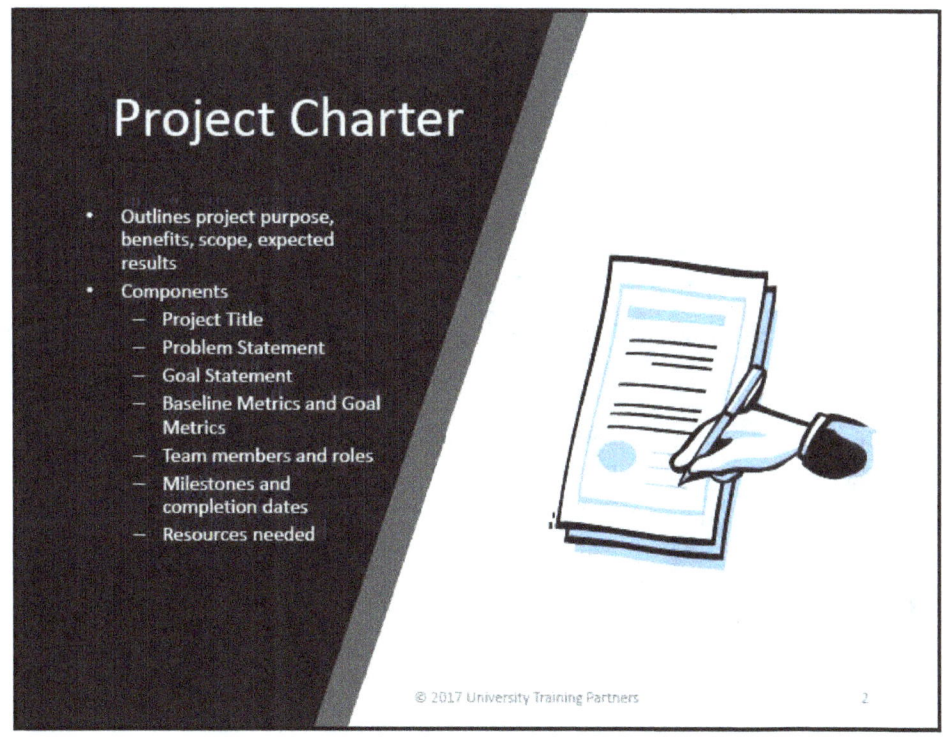

Crafting the Problem Statement

- **Over the past six months**, the number of surface defects on door panels has increased from 5/1000 to 60/1000, resulting in increased rework and warranty claims.

- How long has the problem existed?
- What measurable item is affected?
- What is the performance gap?
- What is the business impact?

Problem Statement Cautions

- Causes of the problem or solutions do not belong in the problem statement
    - That's what the project is for
    - If solution is obvious, then is it really a project?
- Make sure item is measurable
- Need data in the problem statement, not conjecture

Problem Statement Examples

- Customer complaints have always been higher than they should be, resulting in lost sales.

- Reduce sealing defects from 5% to 1% by replacing the sealing machine with an improved model.

- Over the past year, hotel bookings have decreased by 10% due to rude employees, resulting in a $400,000 budget shortfall.

Dr. Deming stated that 94% of all problems, defective goods or services came from the system, not from a careless worker or a defective machine.

What about these?

- Over the past two years, customer dining complaints have increased by 20%, resulting in fewer repeat customers.

- Order fulfillment time for product X has increased from 3 weeks to 12 weeks in the past year, allowing our competitor to increase its market share.

How long
Measurable item
Performance gap
Business impact

How Did You Do?

- **Over the past two years**, customer dining complaints have **increased by 20%**, resulting in fewer repeat customers.

- Order fulfillment time for product X has **increased from 3 weeks to 12 weeks** **in the past year**, allowing our competitor to increase its market share.

How long
Measurable item
Performance gap
Business impact

Goal Statement

- SMART Goal
  - S – Specific
  - M – Measurable
  - A – Achievable
  - R – Realistic
  - T – Time bound

- "Reduce surface defects on doors from 60/1000 to 5/1000 within 6 months."

Building the Gantt Chart

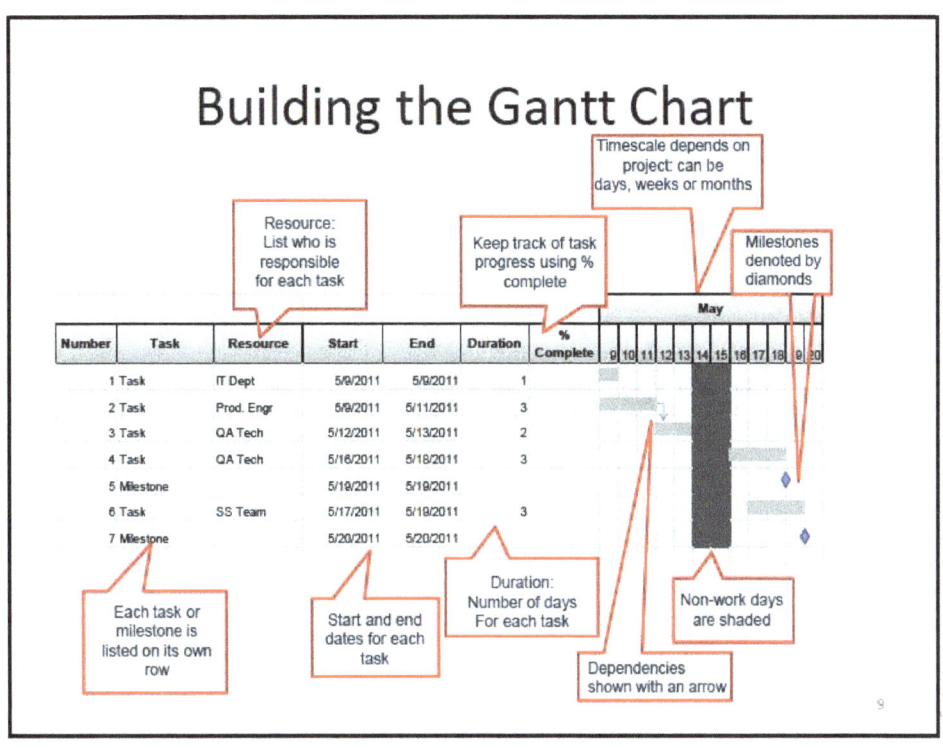

Project Milestone & Tollgates

- Milestones include each phase of the DMAIC process

- Tollgate reviews are held after each phase is completed and before a new phase can begin

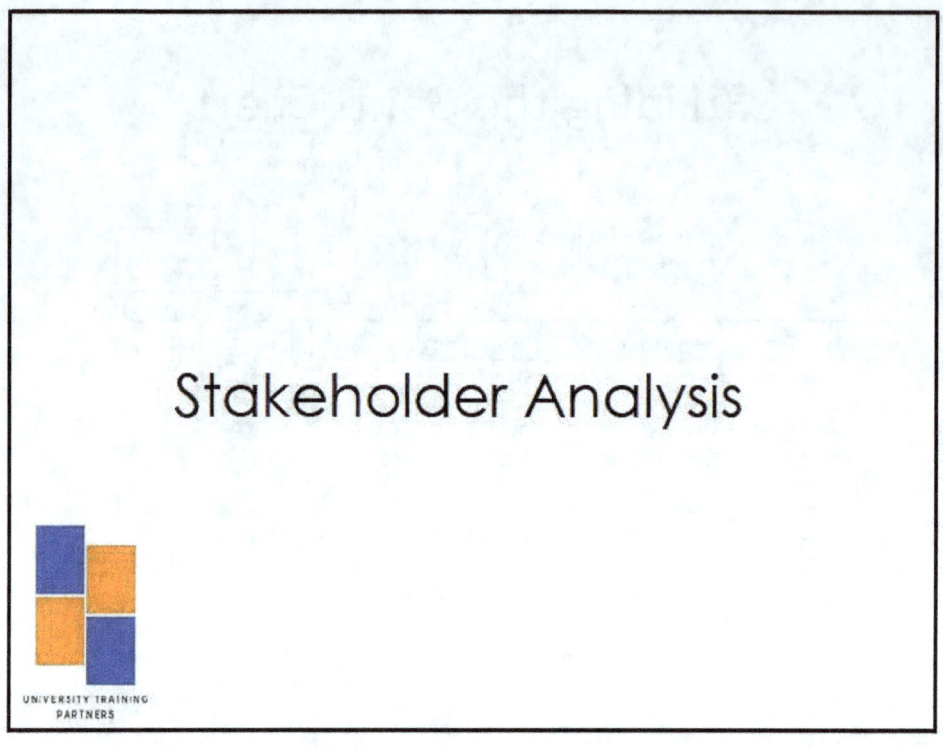

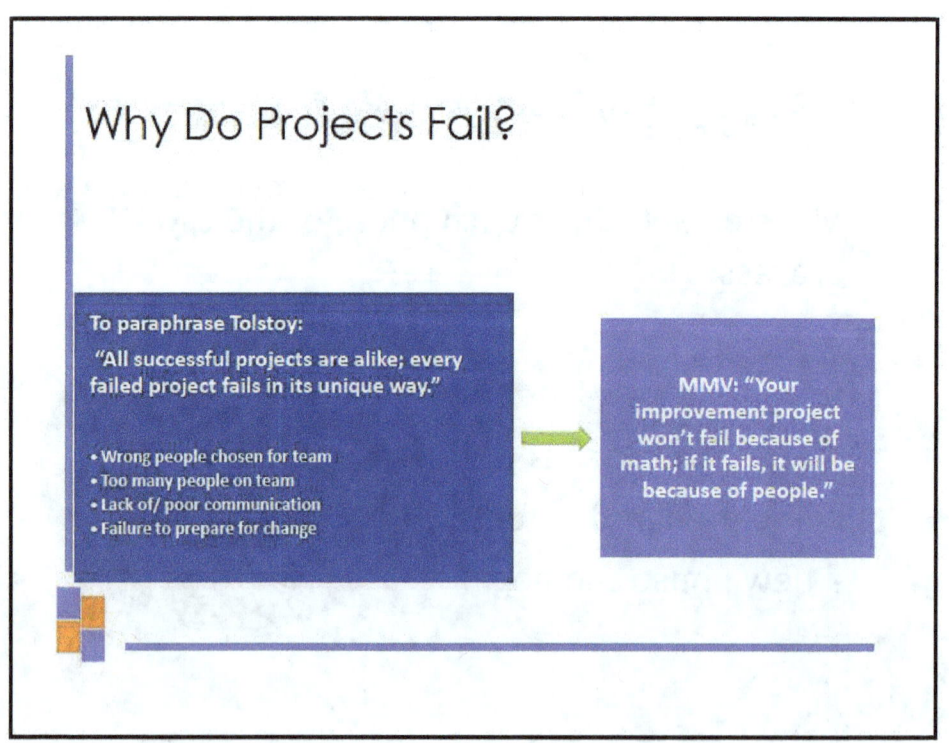

Stakeholder Analysis

1 › 2 › 3 › 4

1. Identify stakeholders (SH)
2. Place each SH on the Influence-Importance grid
3. Select a participation level for each SH
4. Create a communication plan for each SH

Stakeholder Defined

A stakeholder is anyone who has a vested interest in a project or who can effect or be affected by an action or change.

- Who might receive benefits or experience negative effects?
- Who might be forced to make changes or change behavior?
- Who has goals that align/conflict with the project goal?
- Who has responsibility for action or decisions?
- Who has resources or knowledge that are important to the project?
- Who has expectations for this project?

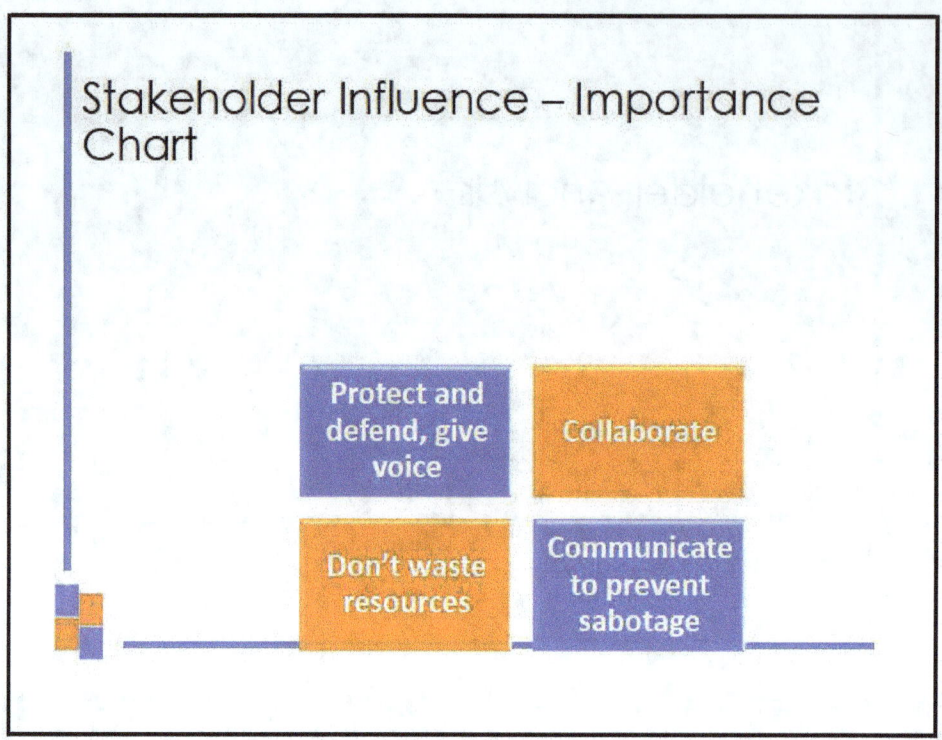
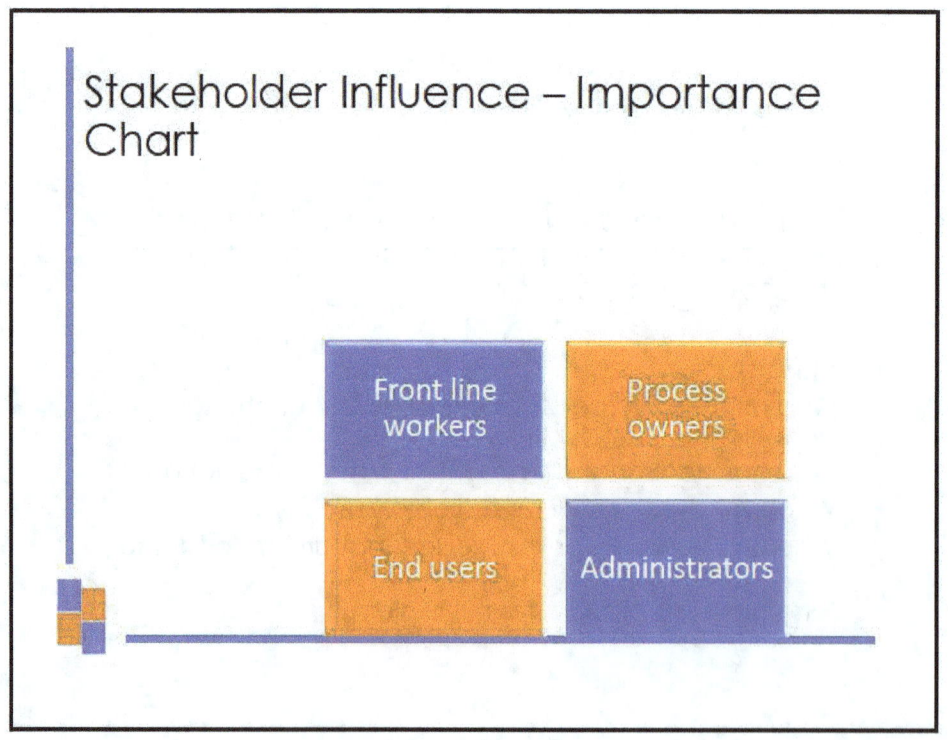

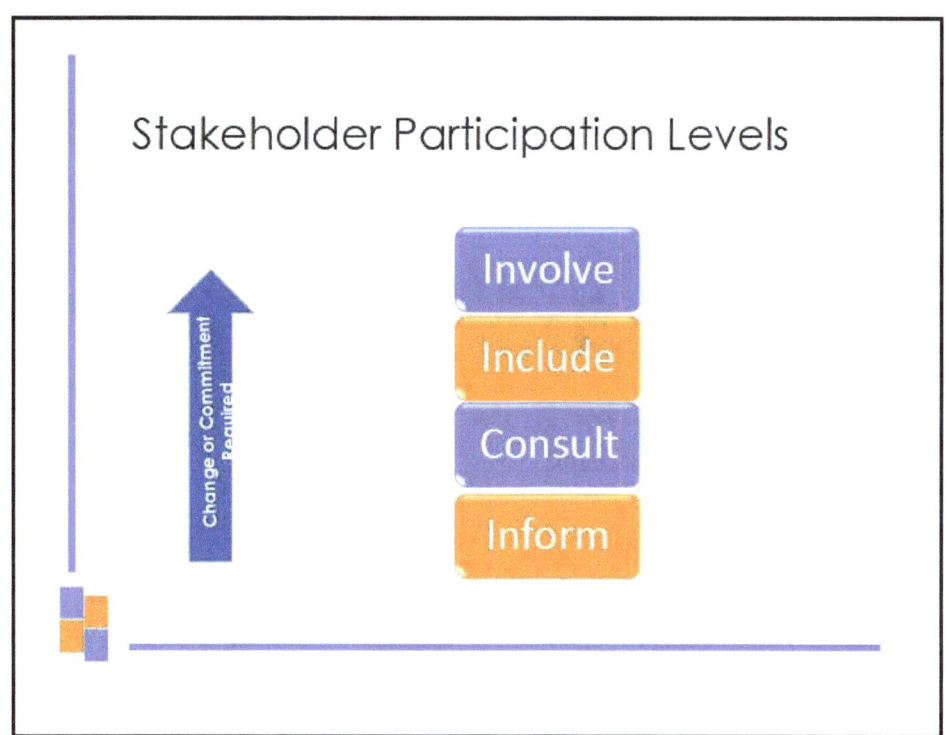

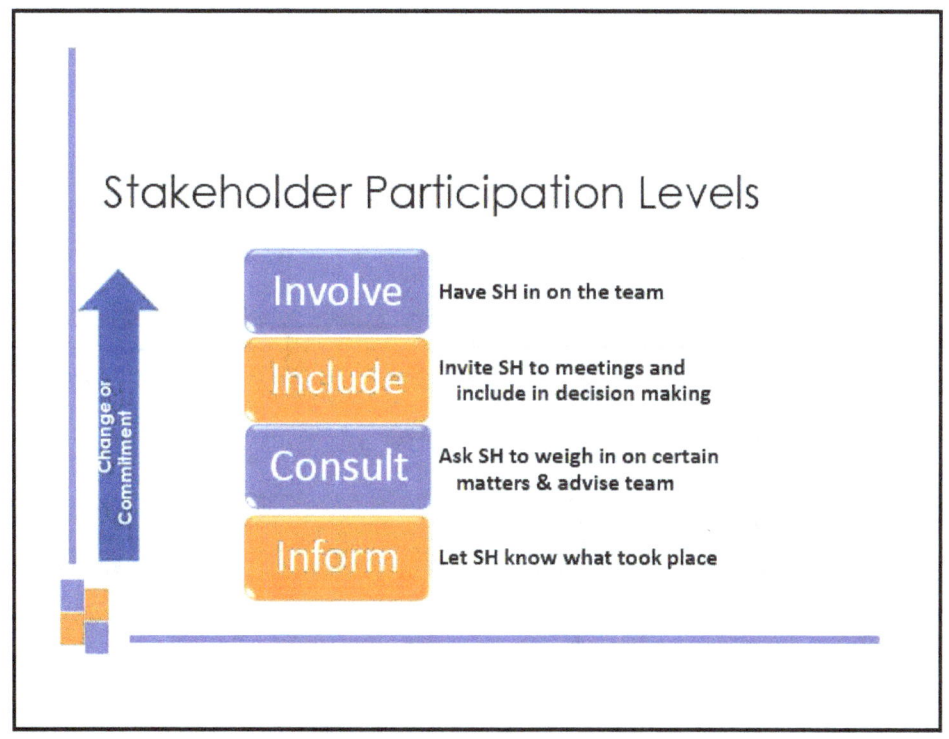

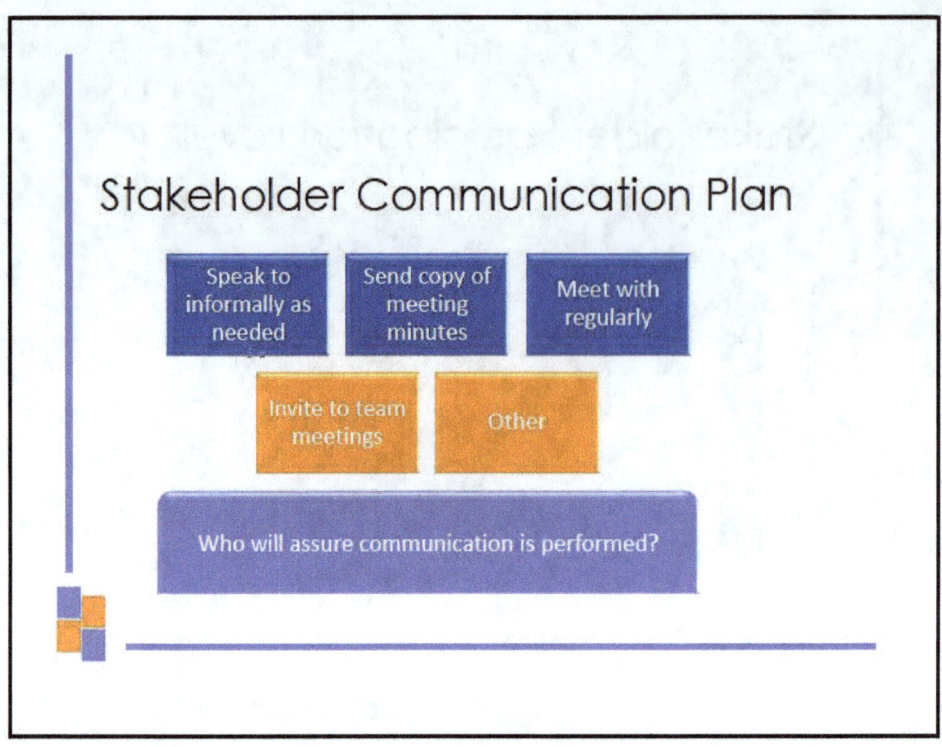

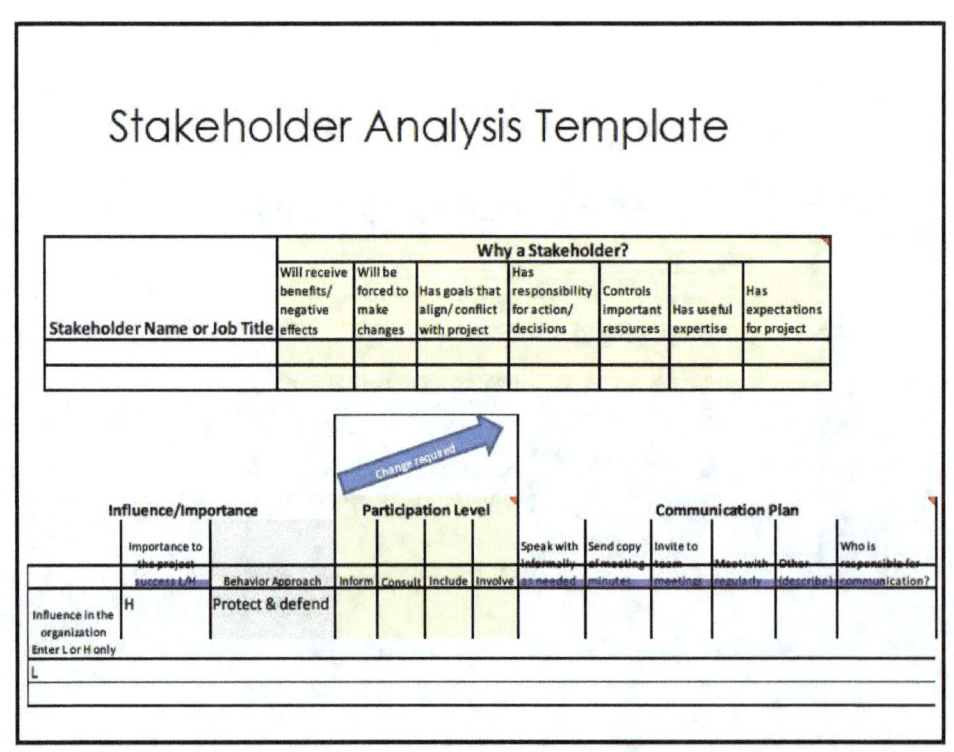

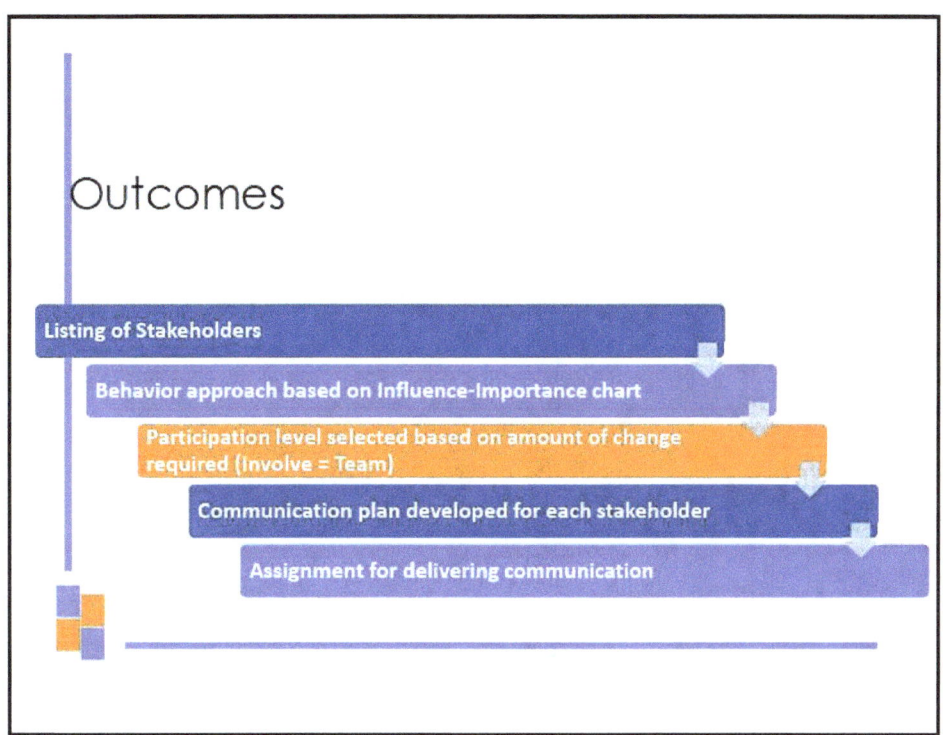

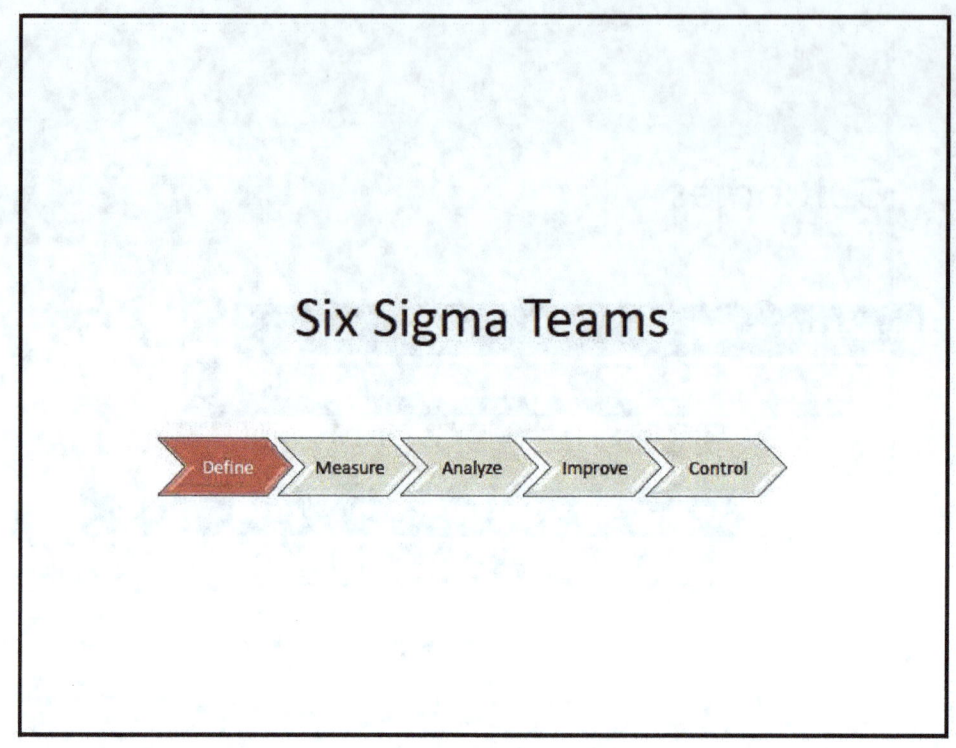

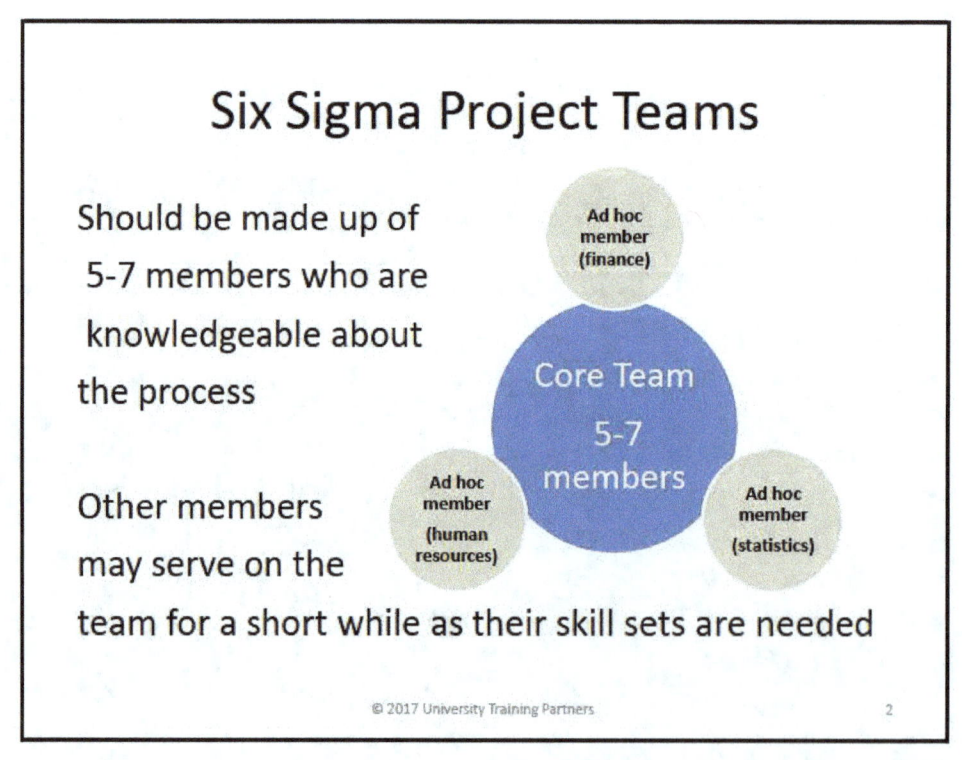

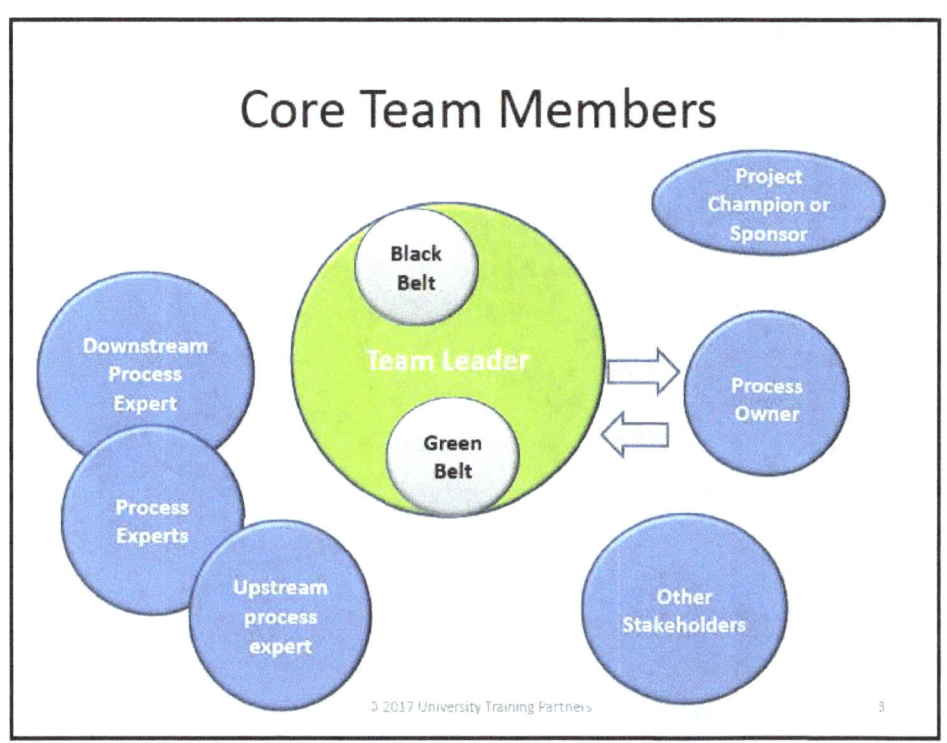

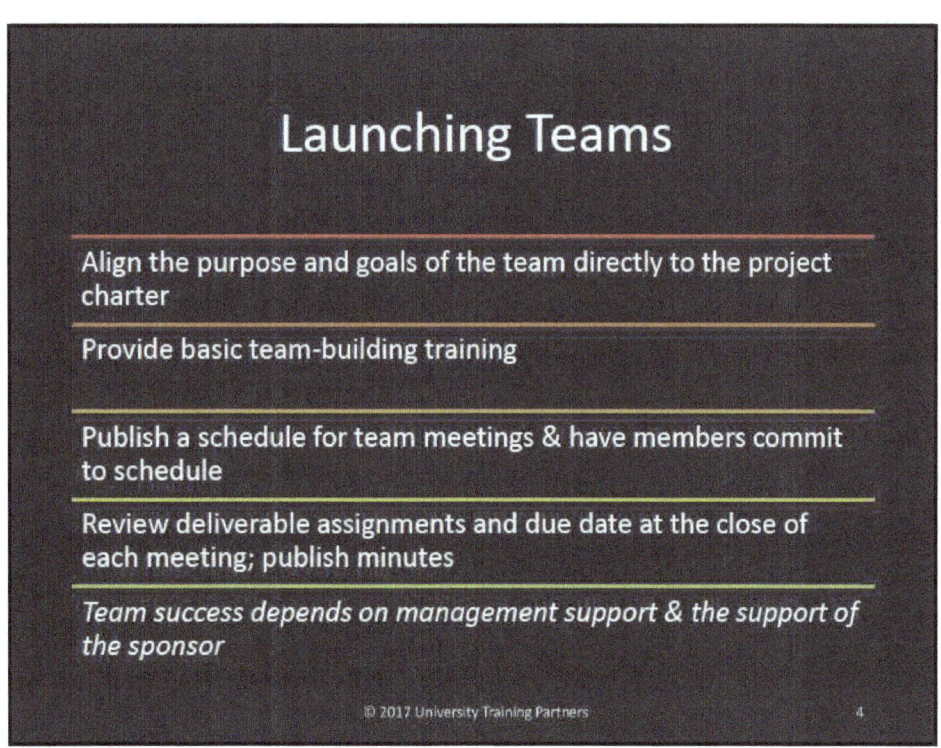

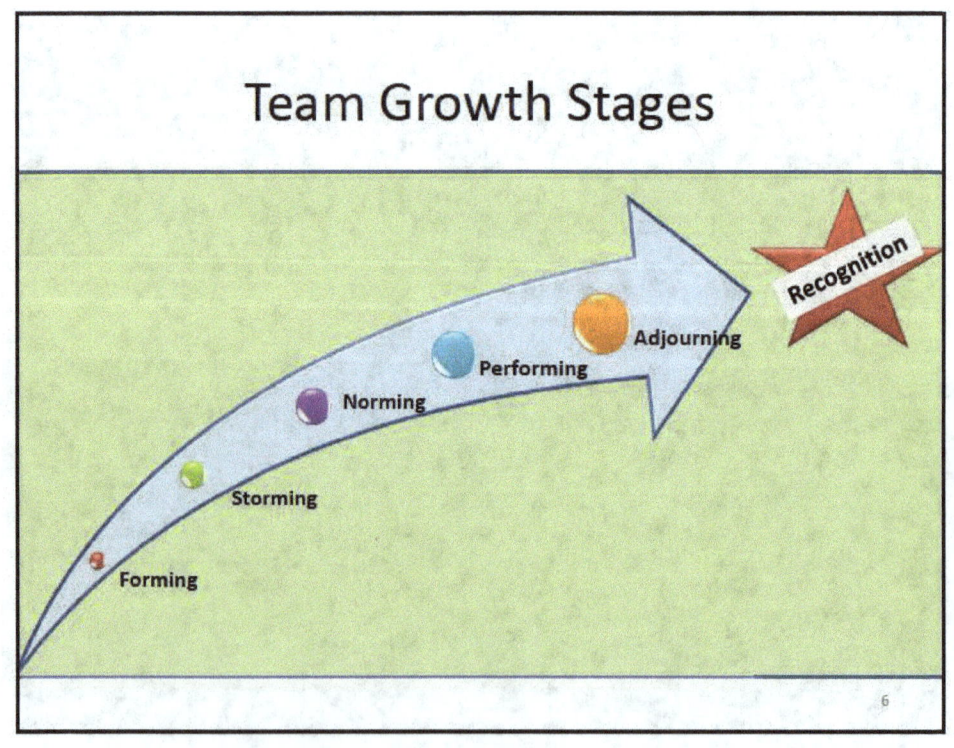

When the team is …

- **Forming**, members struggle to understand the goal and how they will contribute
- **Storming**, members express their own opinions and ideas, often disagreeing
- **Norming**, members begin to understand the need to operate as a <u>team</u>
- **Performing**, team members work together to achieve a common goal
- **Adjourning**, a final meeting is held to discuss & summarize project decisions and wrap up loose ends

Introduction to Process Maps

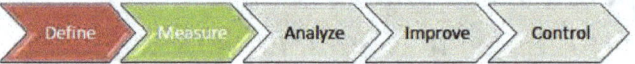

Process Defined

Recall:

A process is a step or a sequence of steps that uses inputs and produces a product or service as an output.

Creating a Process Map

1. Walk the process! **Go to Gemba**
2. Note process boundaries as defined by project
3. Team writes process steps on post-it notes
4. Team arranges steps in sequence
5. Make any necessary adjustments
6. Number tasks sequentially
7. Transfer map to paper or computer

Three Basic Map Symbols

- Start and End Points (oval)
- Activity (rectangle)
- Decision (diamond)

Mapping Tips

- Make sure the level of detail matches your purposes
- Use a cross-functional team to create the map
- Concentrate on documenting the process, not using the right symbols
- Capture ancillary ideas brought up in the map meeting and place in a Parking Lot

Swim Lane Chart

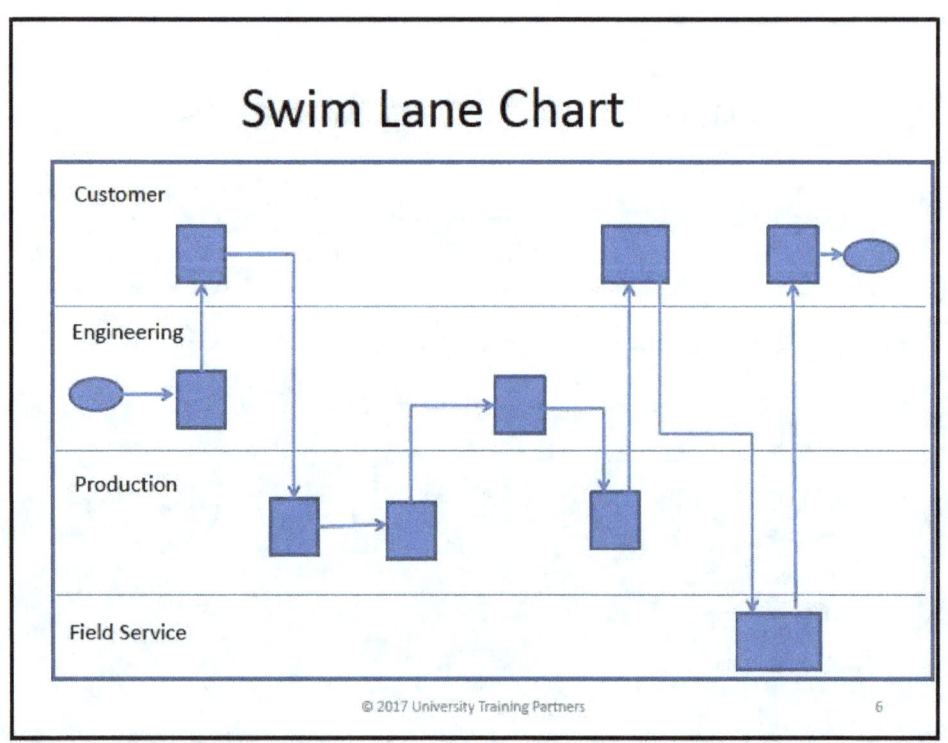

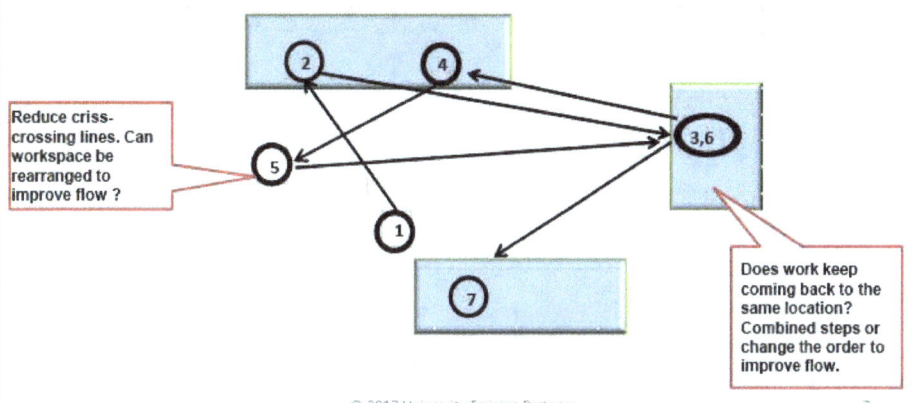

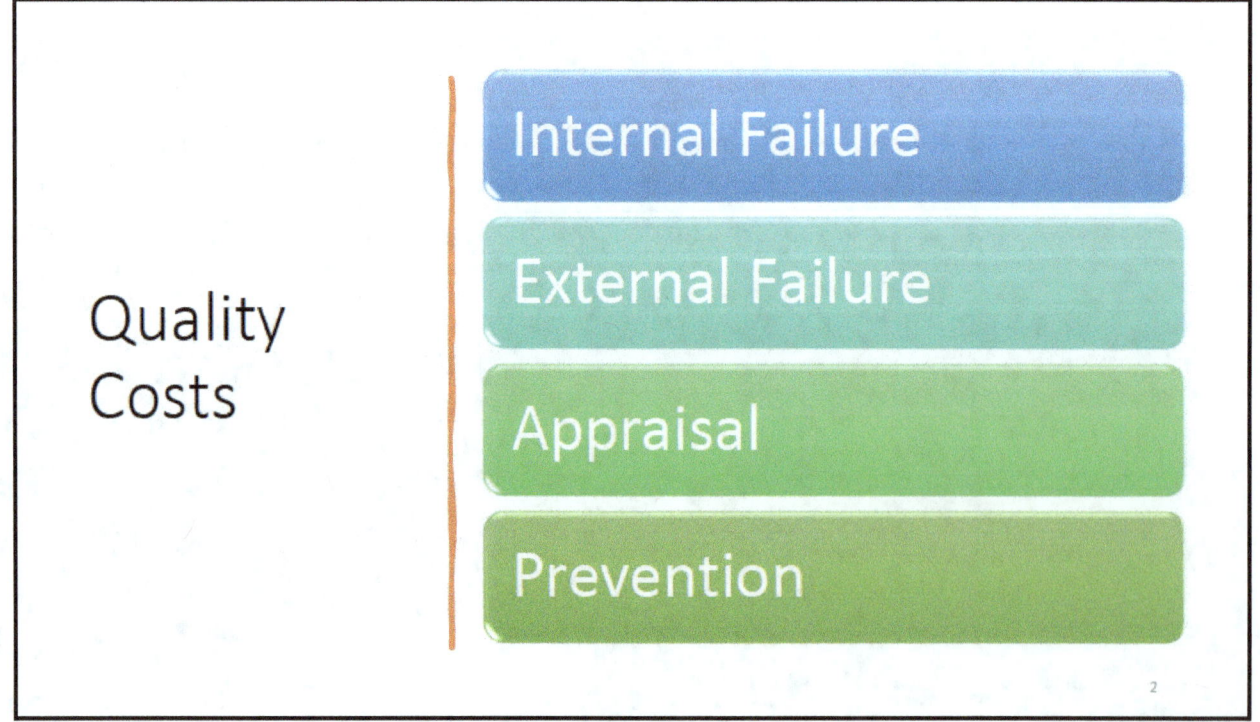

Quality Costs

Internal Failure

Mistakes that the end user doesn't see

- Scrap
- Rework
- Retest
- Downtime
- Yield losses
- Disposition

Quality Costs

Mistakes that the end user sees

External Failure

- Complaint adjustment
- Returned material
- Warranty charges
- Loss of good will

Quality Costs

- Incoming material inspection
- Inspection and test
- Maintaining test equipment
- Materials consumed

Appraisal

Quality Costs

- Training
- Process control
- Mistake proofing
- Visual management

Prevention

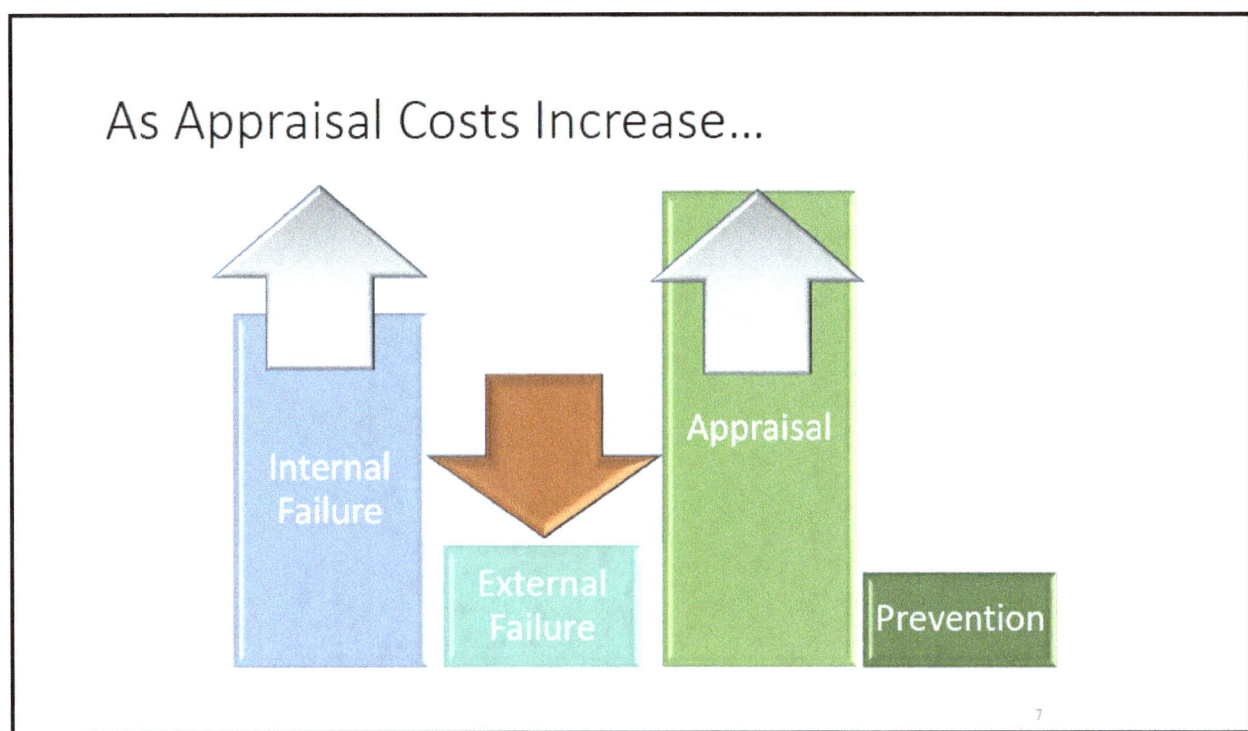

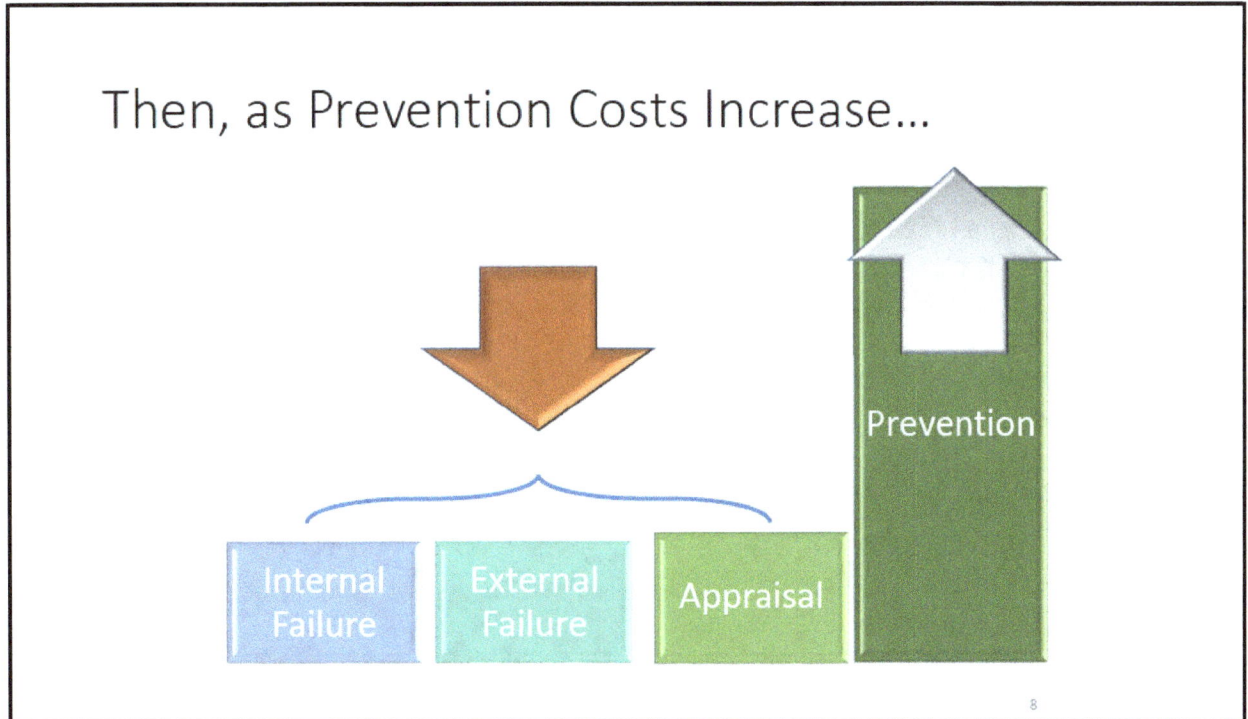

Traditional Quality Cost

Modern Quality Cost

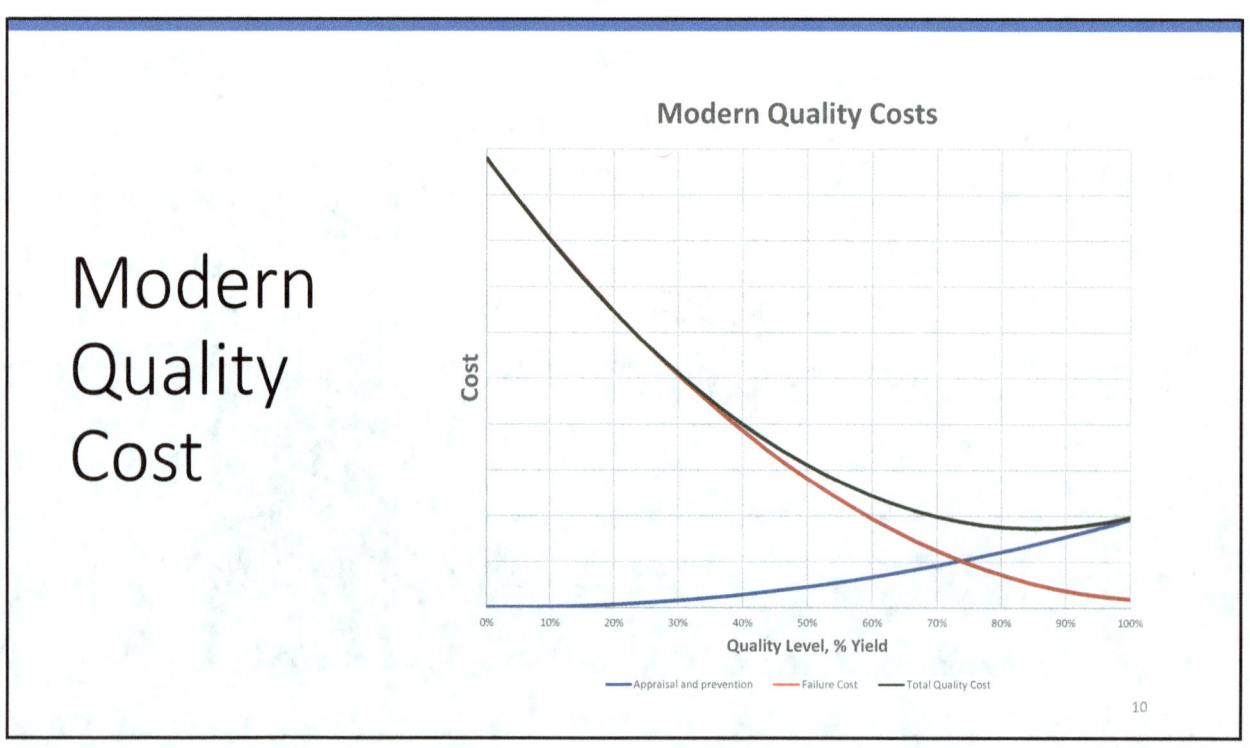

Finding Root Causes

"If we hear hoof beats off in the distance, it is likely to be a horse, not a zebra."

Brainstorming Techniques

- The goal is to generate as many ideas as possible in the time allotted
- Don't evaluate each idea as it is mentioned
- Everything gets written on the board as stated!
- Watch for dominant or intimidating personalities
    - Try a round robin approach
    - Consider individual brainstorming time followed by group sharing

Fishbone Diagram

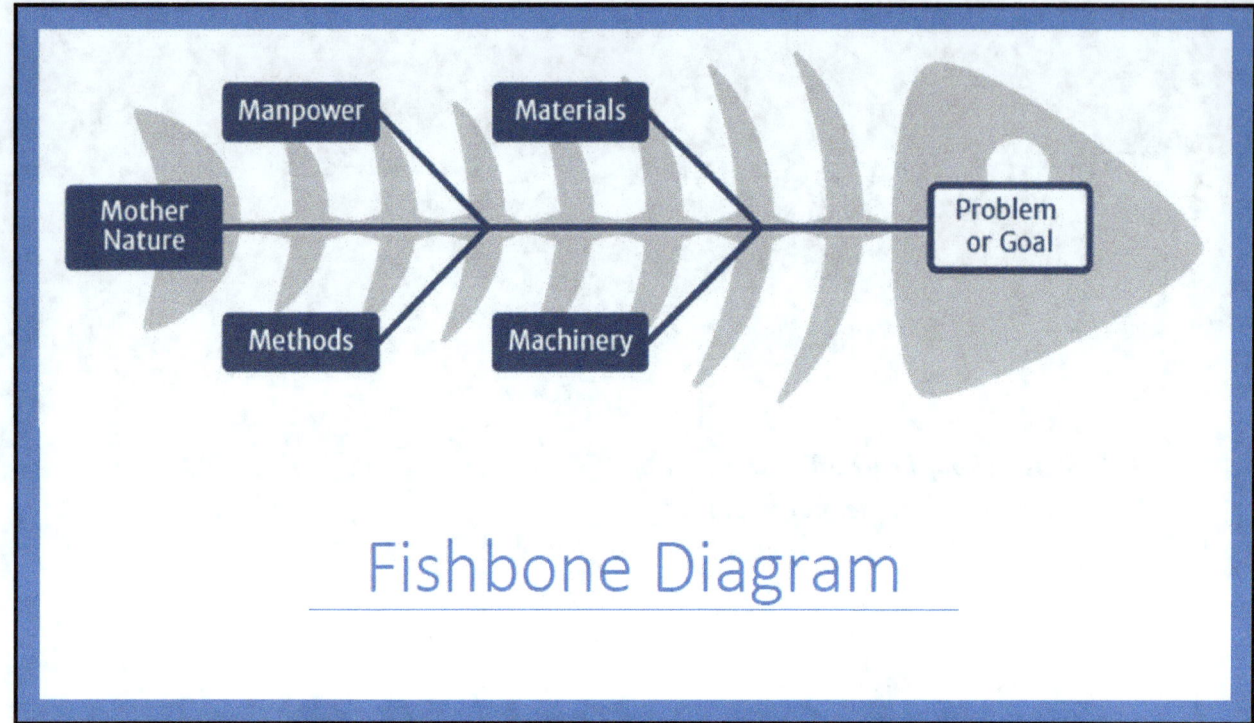

Why-Why Diagram

1. Write the problem or issue on the left-hand side of the white board.

2. Ask "Why?" List all causes on separate sticky notes and place in a column to the right of the problem.

3. Now, treat each of these causes as new problem statements. Ask why and place these causes to the right of the other column. Draw arrows connecting problems and causes.

4. Keep going until the team reaches a fundamental, or system cause, such as company policy.

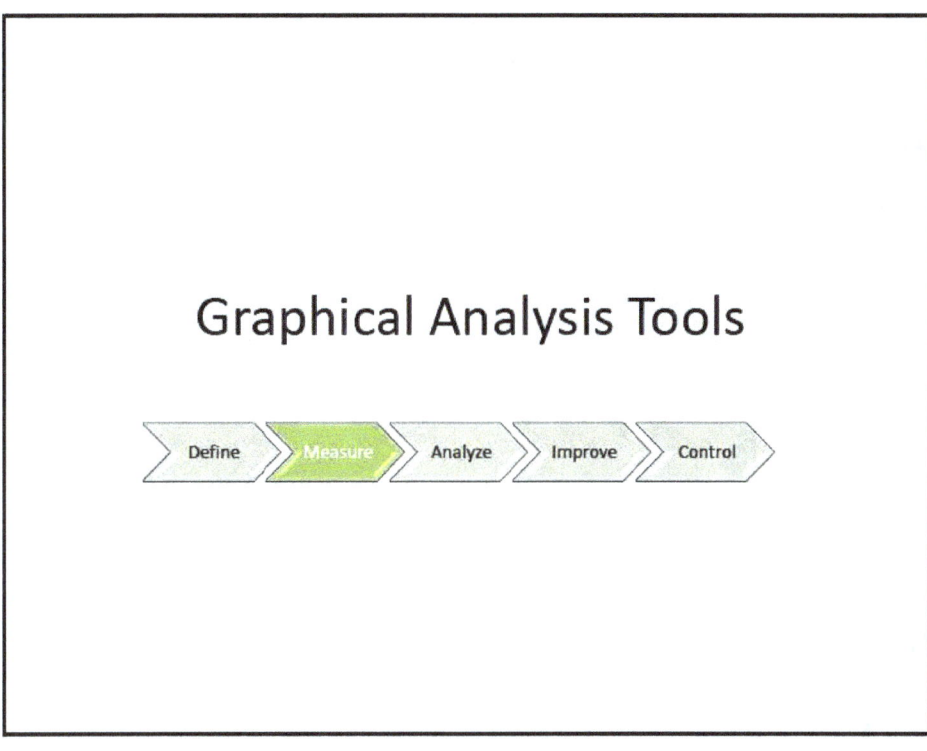

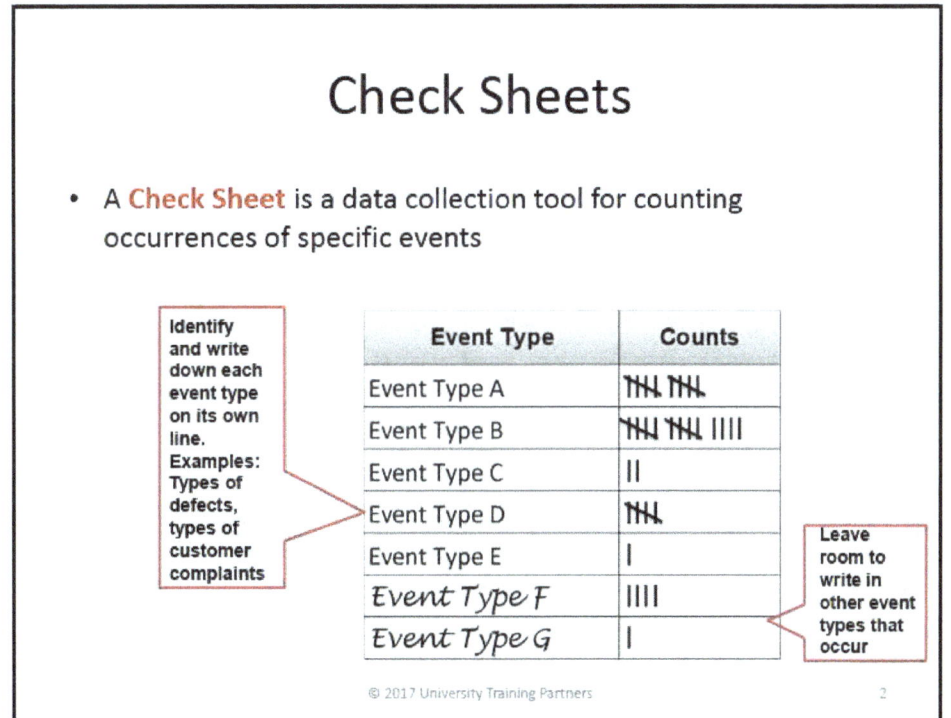

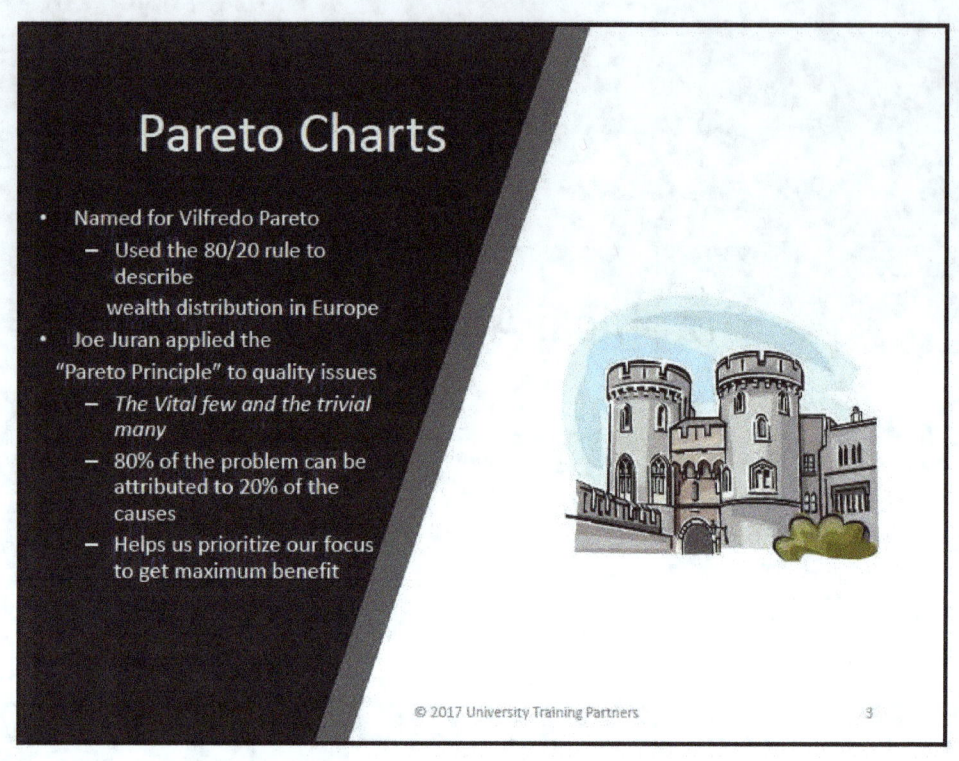

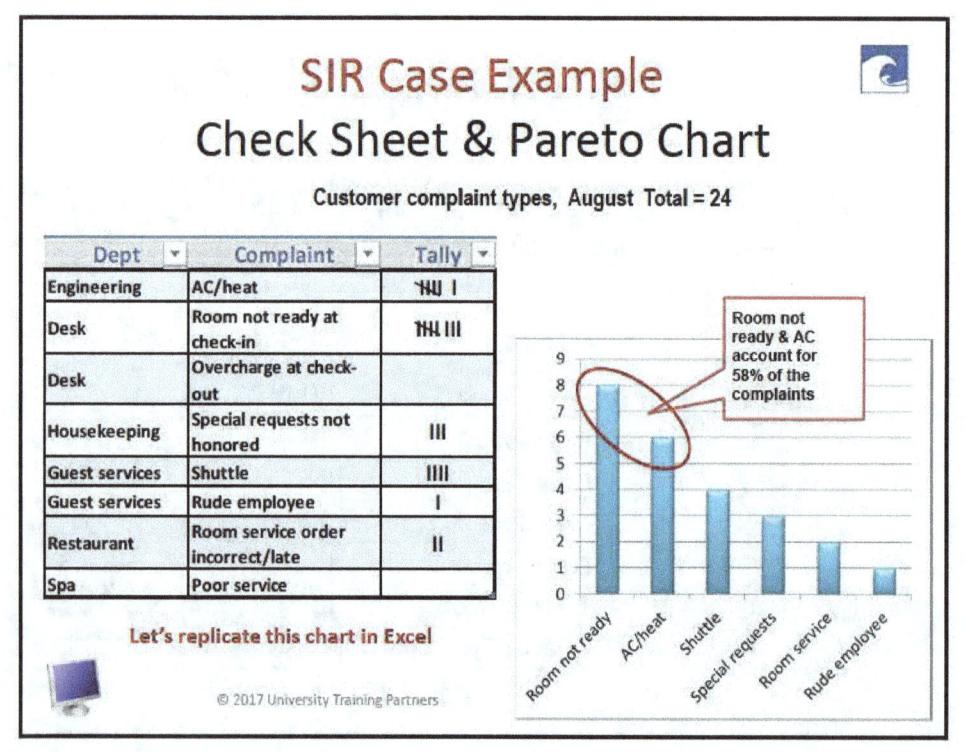

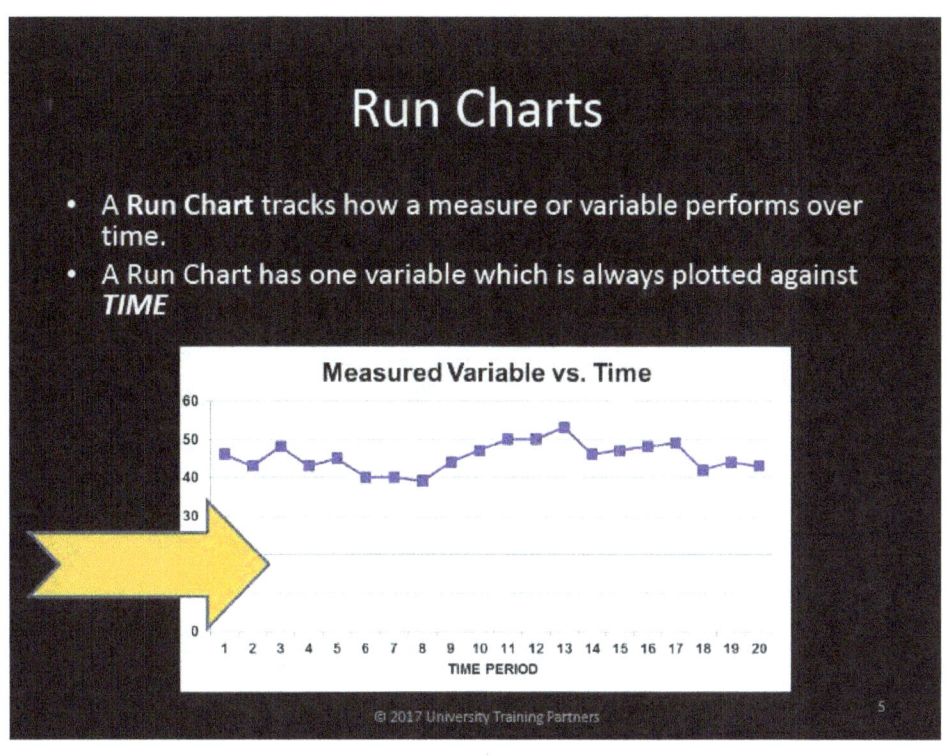

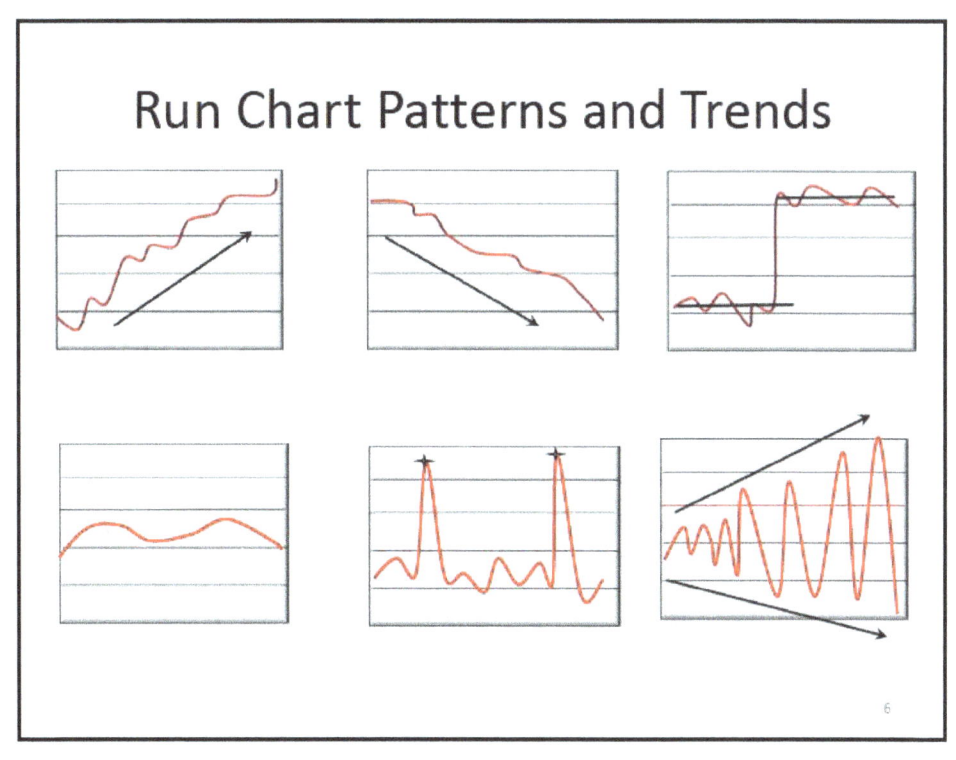

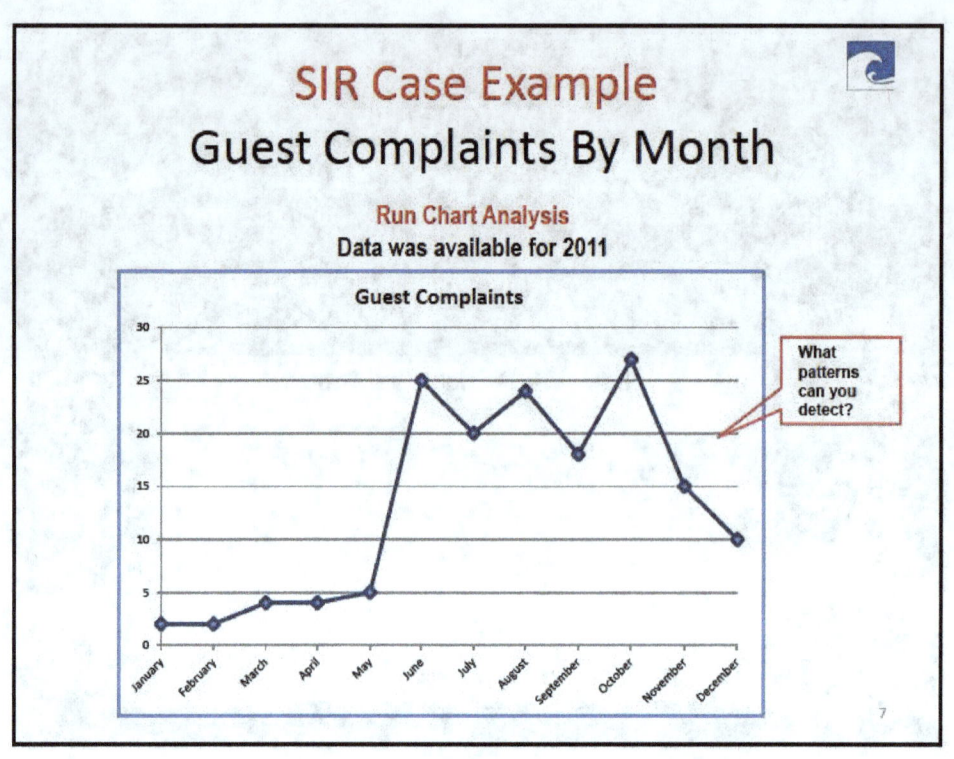
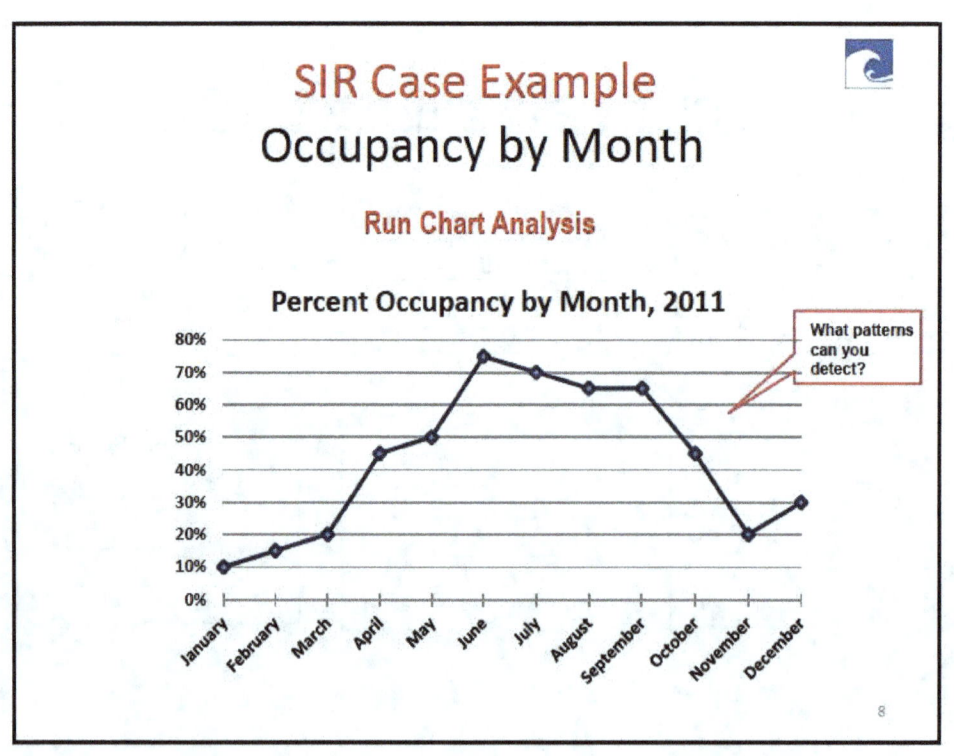

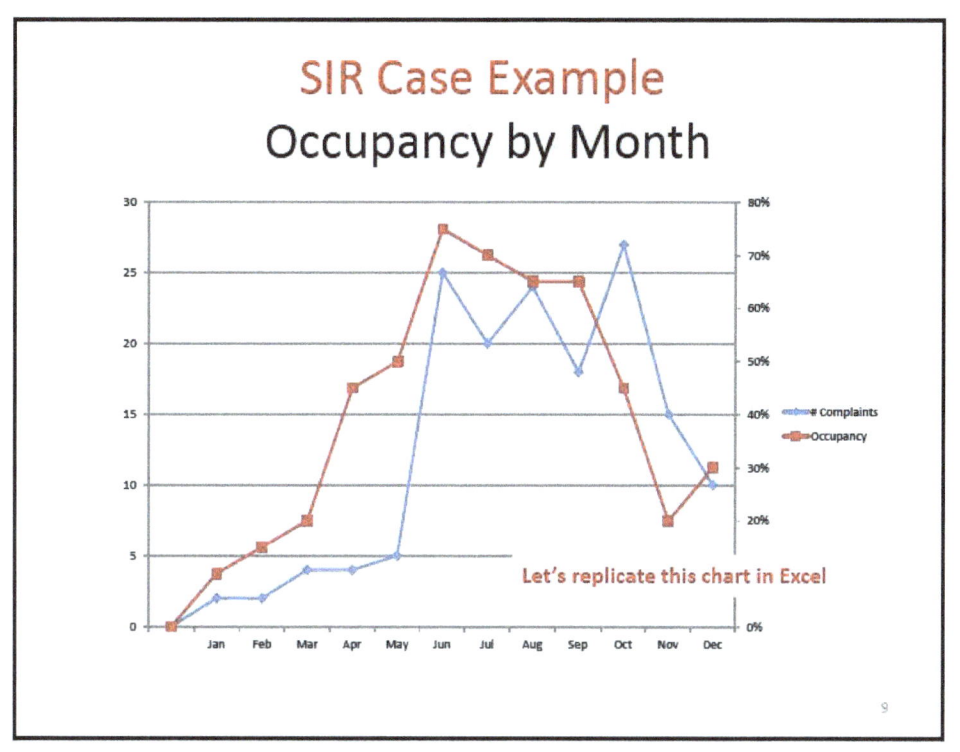

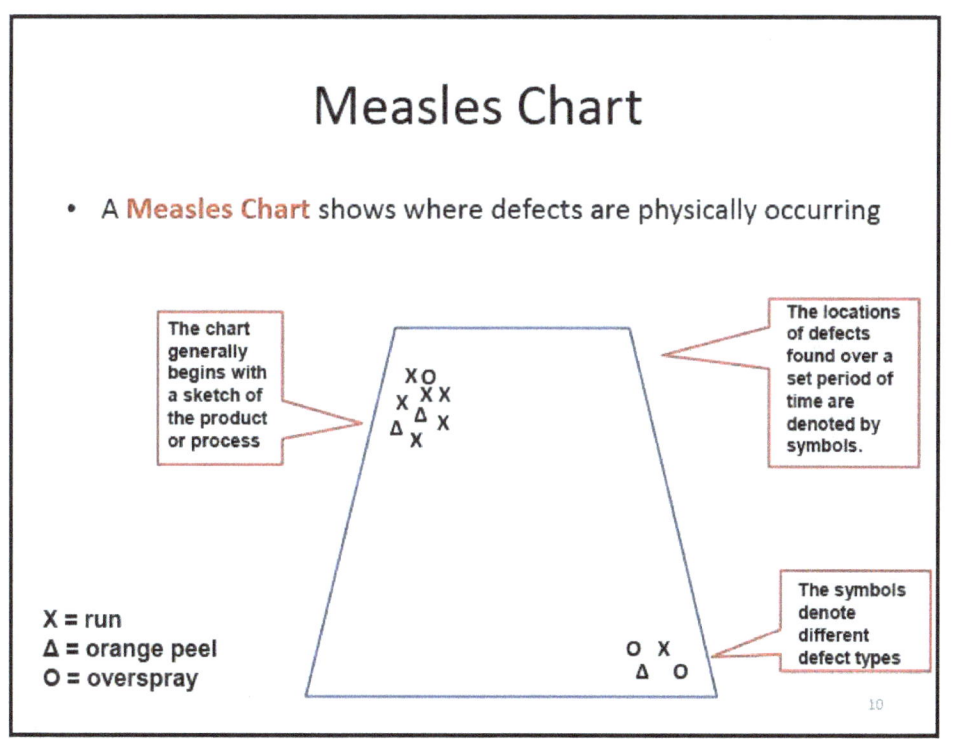

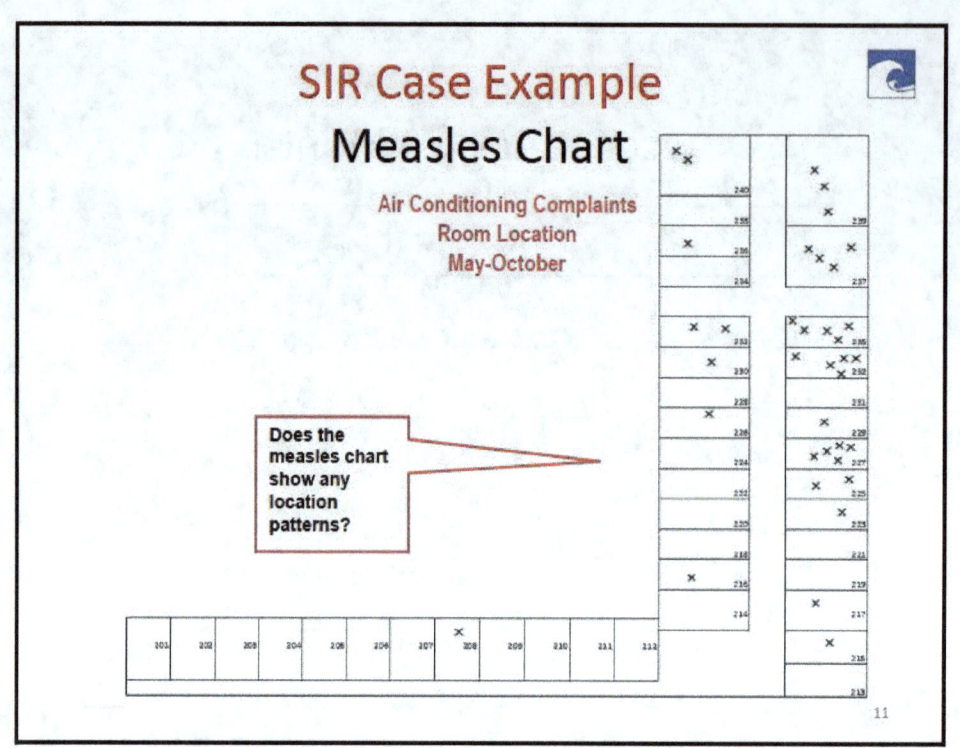

More Graphical Analysis Tools

Frequency Table

Test score raw data:
53, 55, 71, 72, 76, 82, 82, 85, 86, 89, 93, 94, 98, 100

Grade Range	Count
90-100	4
80-89	5
70-79	3
60-69	0
50-59	2

- Categories make "business sense"
- Categories include all data
- There is no overlap between categories
- Each data point fits into one and only one category
- Category ranges are the same size

Do you notice an exception to one of the guidelines here?

Histograms

- Graphical display of a frequency table
- Give a sense of the shape of the data distribution

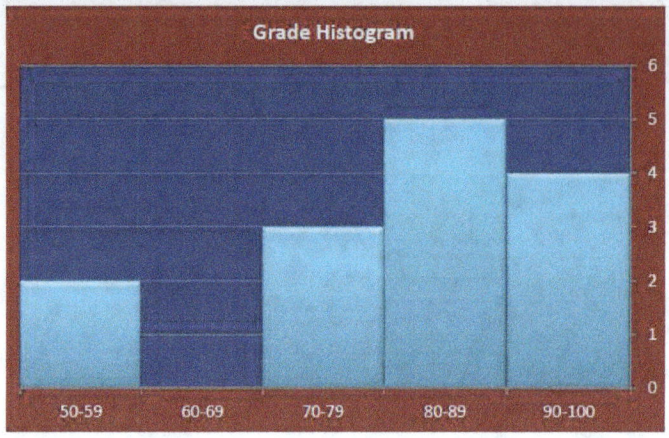

Compare Data Displays

Test score raw data: 53, 55, 71, 72, 76, 82, 82, 85, 86, 89, 93, 94, 98, 100

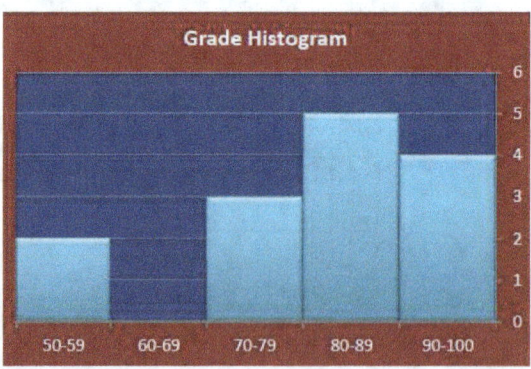

How Many Categories Should We Use?

Too few **Too many**

- Aim for k = 5 to 20 categories
- Can use a rule of thumb: $k \cong \sqrt{n}$

Mirasol Wait Times

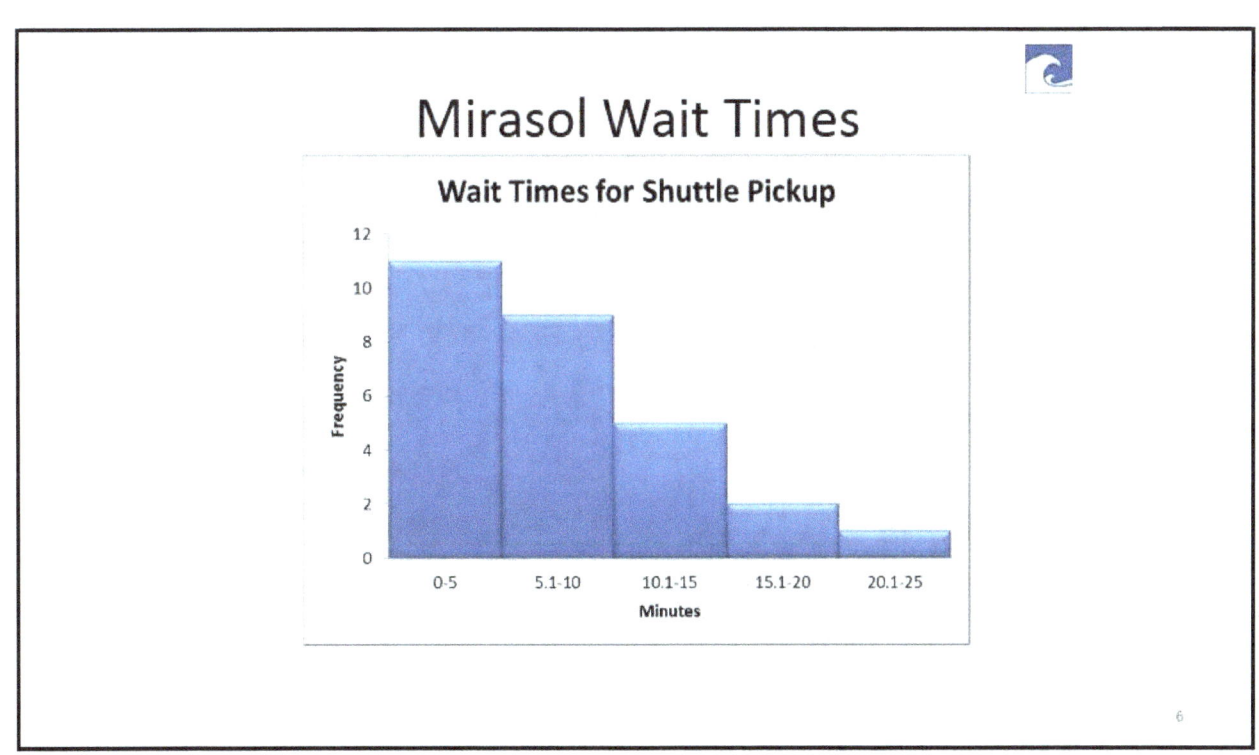

Scatter Plots

- A **Scatter Plot** displays the relationship between two variables

  – Height and weight
  – Home price and square footage
  – Room occupancy and revenue

SIR Case Example
Avg Length of Stay vs. Revenue

Data was available for the last 18 months

Is this too little data? Should we ignore these results?

Scatter Diagram Analysis

Month	Avg Length of Stay	Revenue
1	1.9	$ 90,500
2	4.6	$ 55,400
3	2.5	$ 100,500
4	2.4	$ 125,400
5	2.2	$ 175,400
6	3.2	$ 490,000
7	1.6	$ 55,800
8	1.7	$ 75,600
9	2.8	$ 130,200
10	3.6	$ 305,100
11	2.5	$ 372,000
12	3.5	$ 697,500
13	3.9	$ 694,400
14	3.6	$ 644,800
15	3.7	$ 487,500
16	1.7	$ 292,950
17	1.5	$ 96,000
18	2.6	$ 195,300

Monthly Revenue vs. Avg Length of Stay

What type of relationship seems to exist?

What can be said about this point?

Scatter Diagram Cautions

- Don't confuse correlation with causality
    - Shark attacks and ice cream sales

- Don't extrapolate
    - *"Past performance does not guarantee future results."*

Making Sense of Data

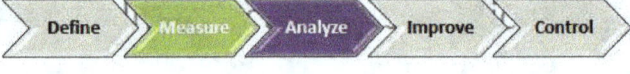

Types of Data

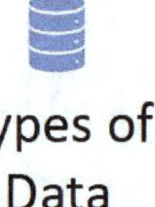

- **Quantitative data** consists of numbers that are counts or measurements

- **Qualitative data** can be separated into different categories based on non-numeric criteria (can't be added or subtracted meaningfully)

Types of Quantitative Data

- **Continuous data** is measured on a scale that can be infinitely divided
  - Length, weight, time, pressure, volume, etc.

- **Discrete or Attributes data** includes counts, binary data and ordinal data
  - Complaint counts
  - Go/no go checks

Statistical Analysis

- **Descriptive Statistics** are used to characterize a data set or population

- **Inferential Statistics** are used to draw conclusions about a population based on analysis of a sample

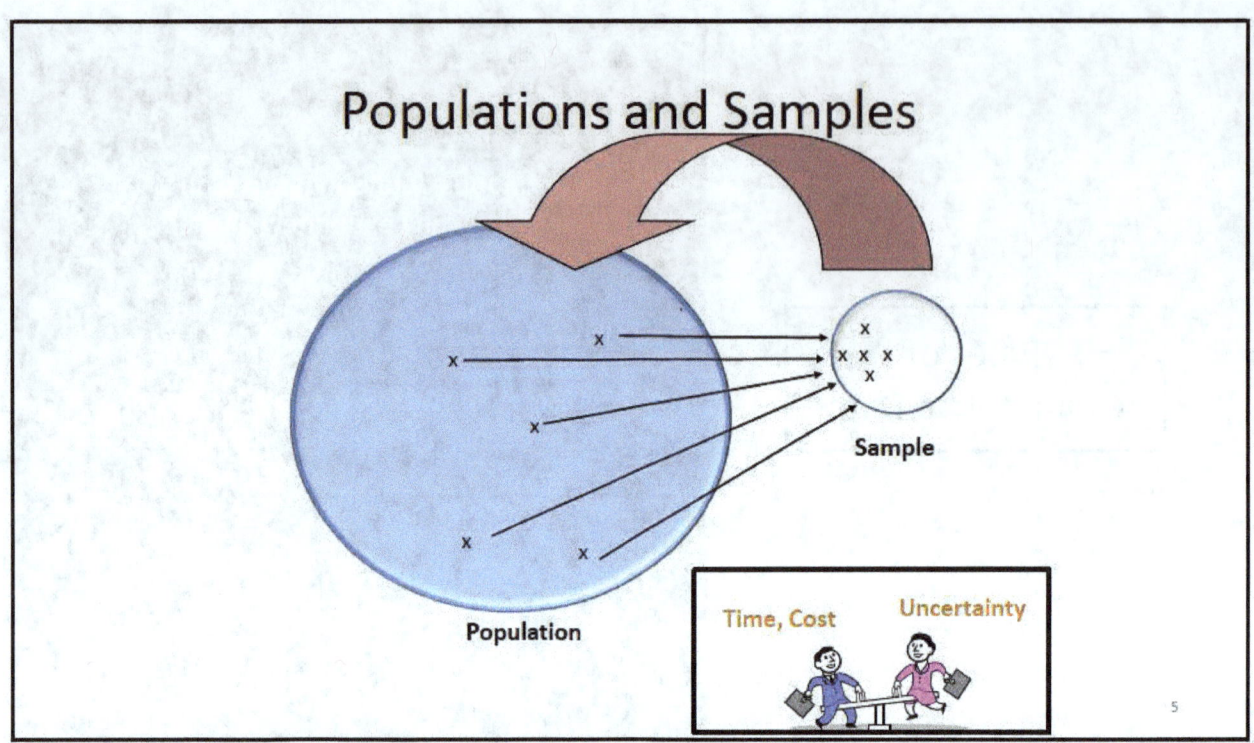

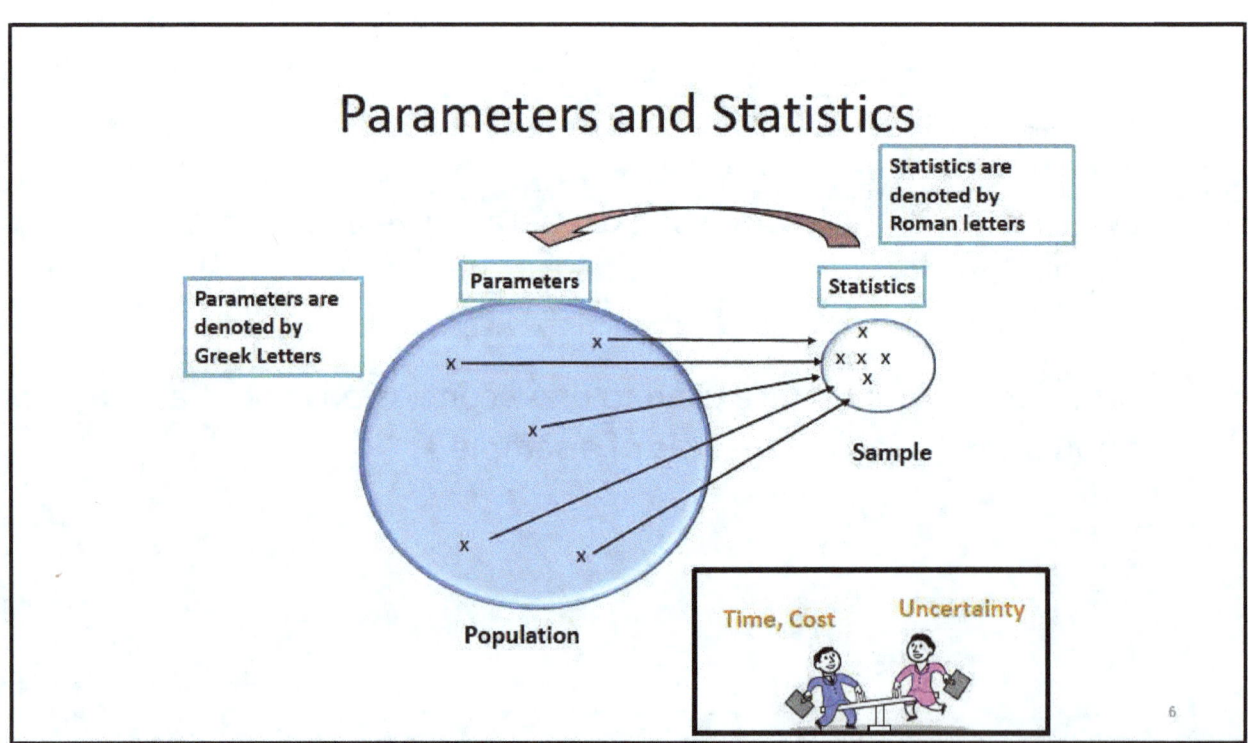

Central Tendency and Dispersion

Central Tendency describes the location of a data set.

Dispersion describes the spread of the data values.

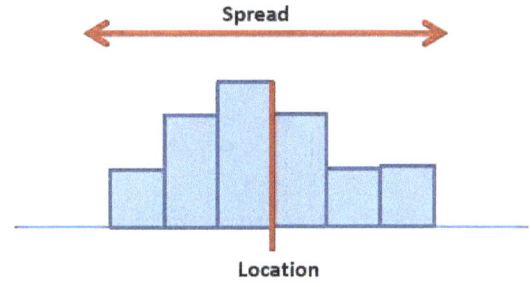

Measures of Central Tendency & Dispersion

Mean

The **mean** is the arithmetic average of a data set

$$\bar{x} = \frac{\sum x}{n}$$

- X bar
- Sum up the data
- Divide by the number of data points

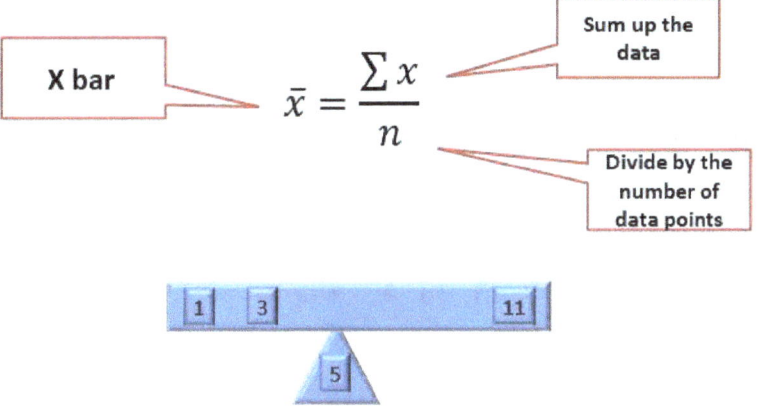

Median

The **median** is the middle of a data set, $\tilde{x}$

1. Sort the data from low to high
2. If n is odd,

$$3, 5, \boxed{7}, 12, 13$$

3. If n is even,

$$3, 5, 7, 12, 13, 15$$

$$\tilde{x} = (7 + 12)/2 = 9.5$$

Mode

The **mode** is the most frequently occurring value in a data set, $\dot{x}$

1. Sort the data from low to high
2. Mode is the most frequently occurring value

$$34, 45, \boxed{46, 46,} 48, 49, 52$$

More on the Mode

23, 25, 26, 28, 29, 30, 32

65, 67, 67, 69, 70, 70, 83

Find the mode of each data set

Features of the Statistics

- Mean, $\bar{x}$
  - Uses all data values in the data set
  - Can be influenced by outliers
- Median, $\tilde{x}$
  - Not influenced by outliers
- Mode, $\dot{x}$
  - Not influenced by outliers
  - May not exist
  - May have multiple values

Central Tendency and Shape

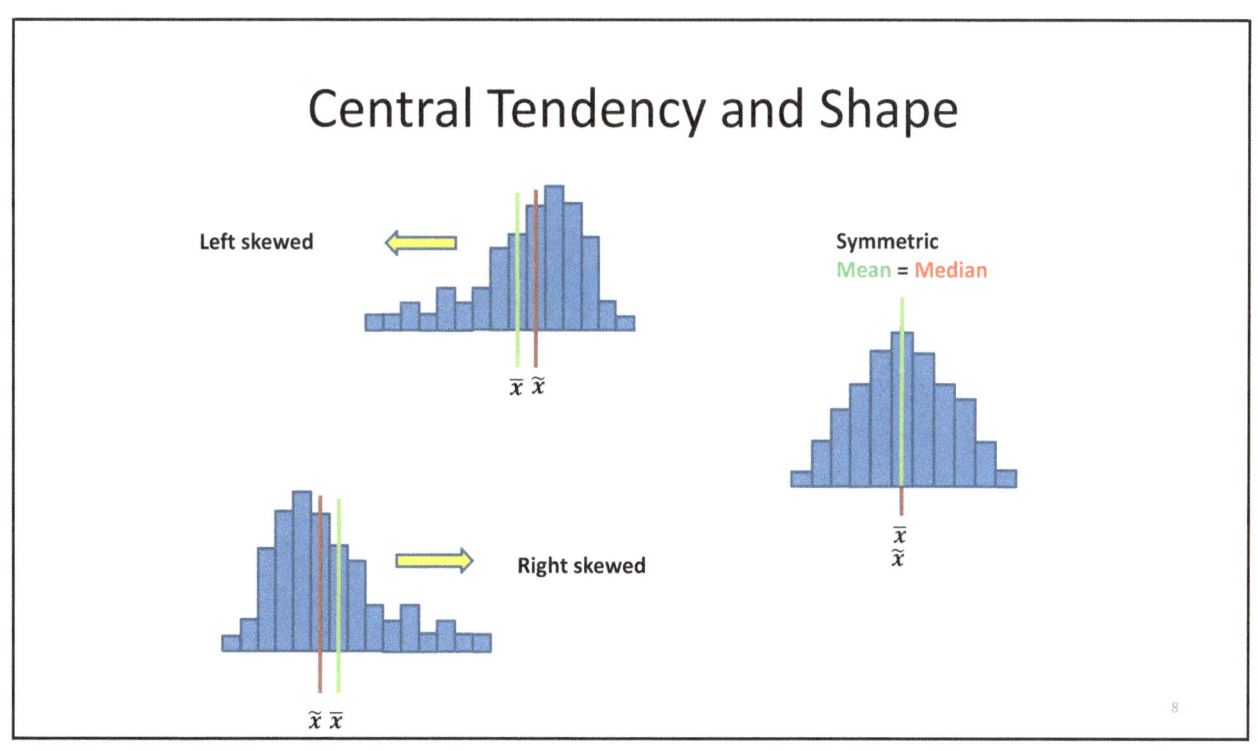

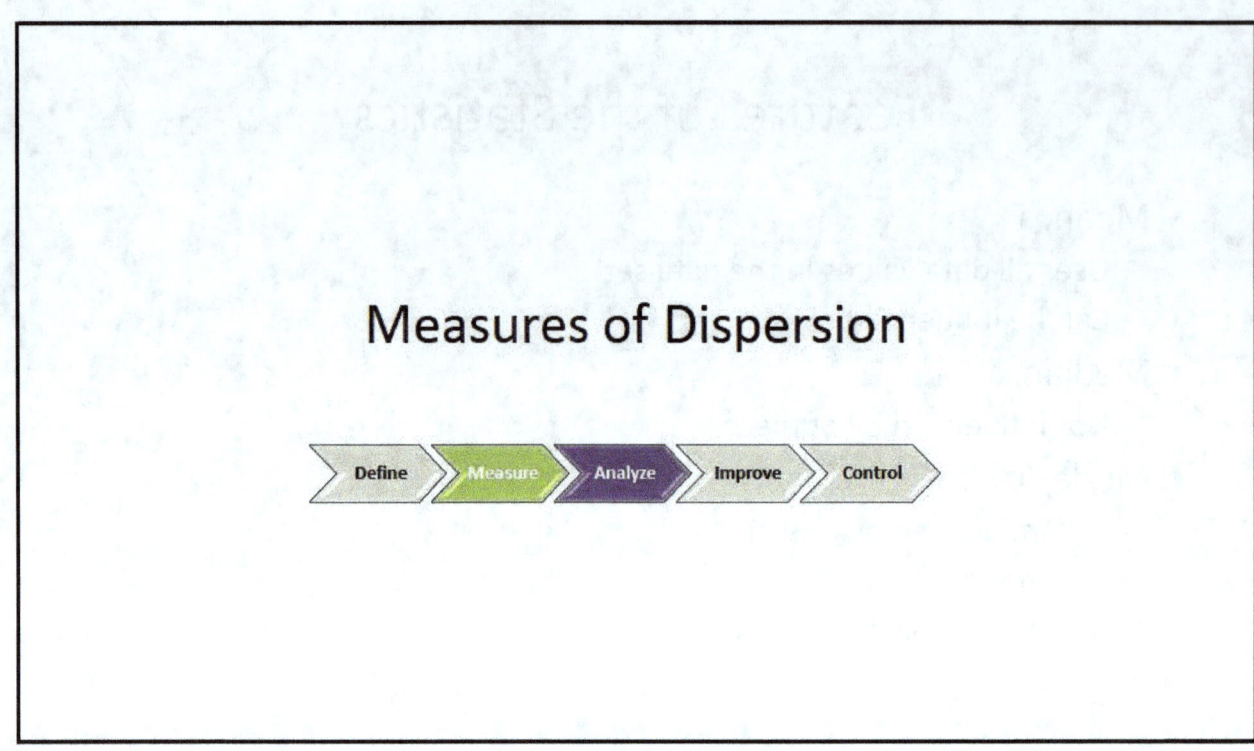

Range and Standard Deviation

The **range** is the difference between the largest data value and the smallest data value

$$R = (x_{max} - x_{min})$$

The **sample standard deviation** is a measure of the spread of the data

$$s = \sqrt{\frac{\sum(x-\bar{x})^2}{n-1}} \qquad \boxed{s = \sqrt{\frac{\sum x^2 - n\bar{x}^2}{n-1}}}$$

Excel Approach

$$= stdev(6,8,8,9,14)$$

Highly recommended!

Features of the Range

- The range only depends on the smallest and largest data values in the sample
- Very sensitive to <u>outliers</u>

6, 8, 8, 9, 14

6, 8, 12, 14, 14

6, 13, 13, 13, 14

What are the ranges of these data sets?

What can we say about the standard deviations of these data sets?

6, 13, 13, 13, 100   What is the range?

Location and Dispersion

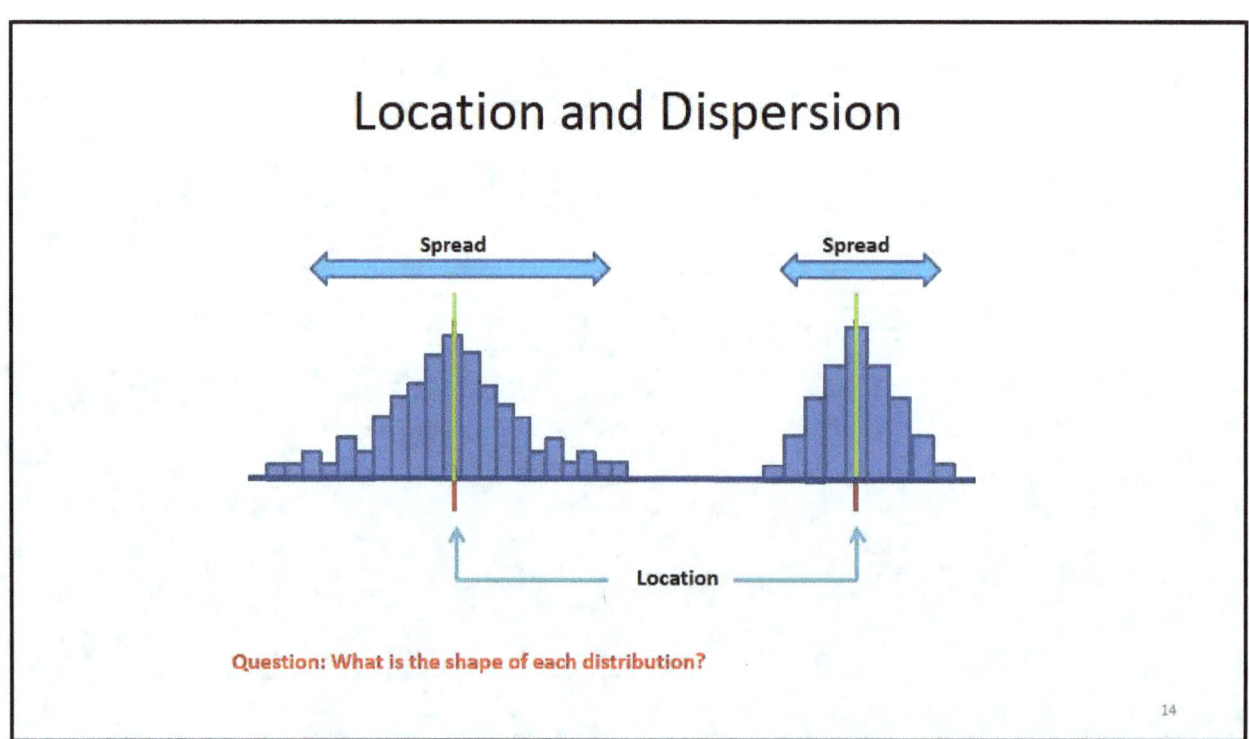

Question: What is the shape of each distribution?

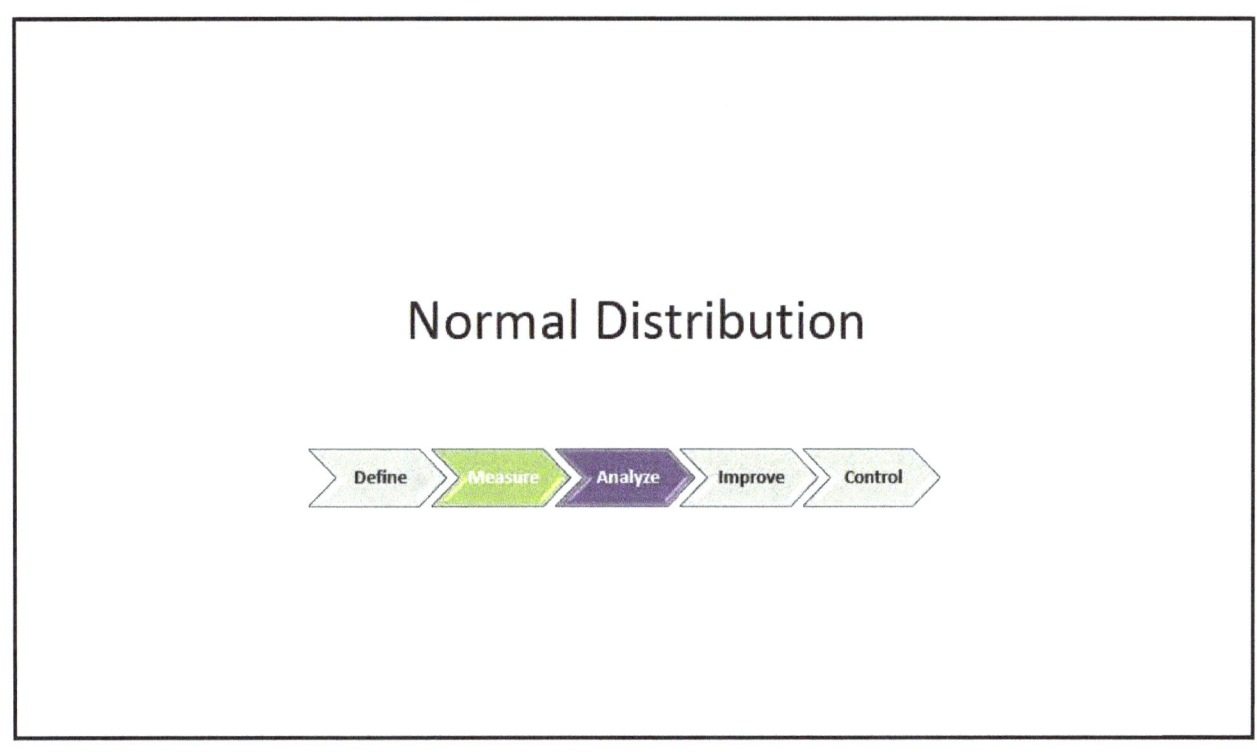

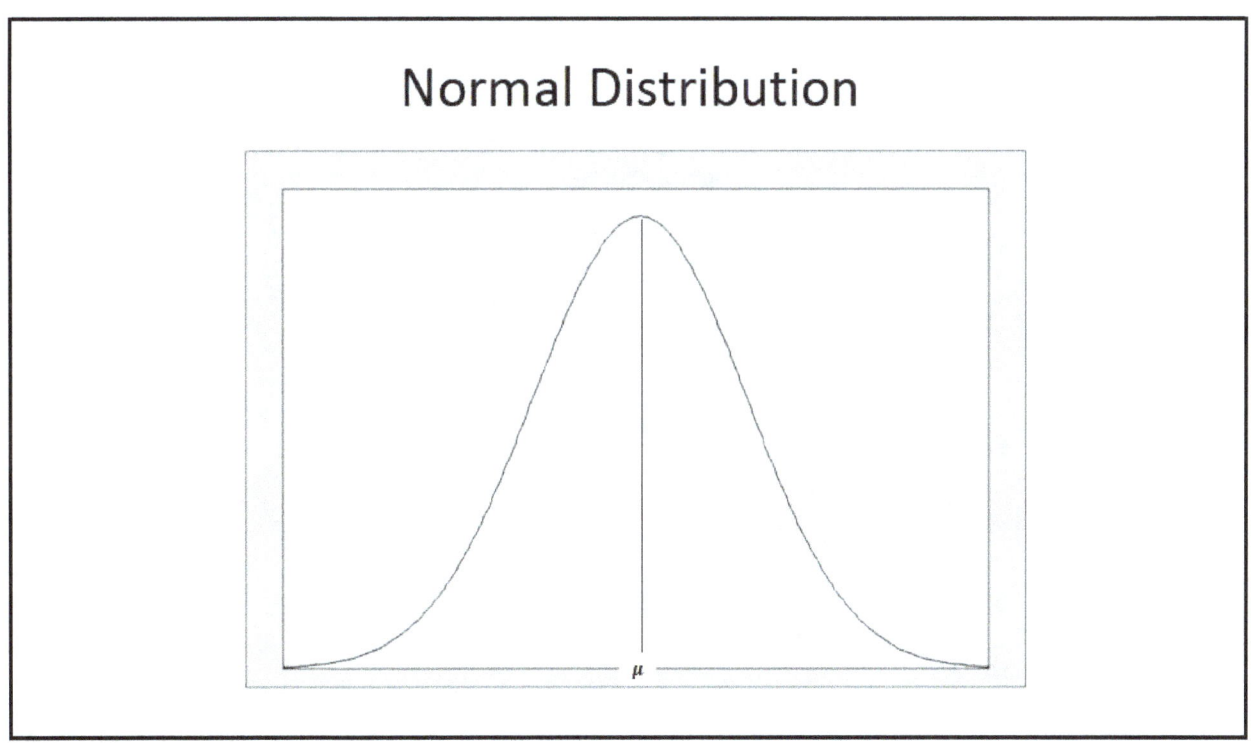

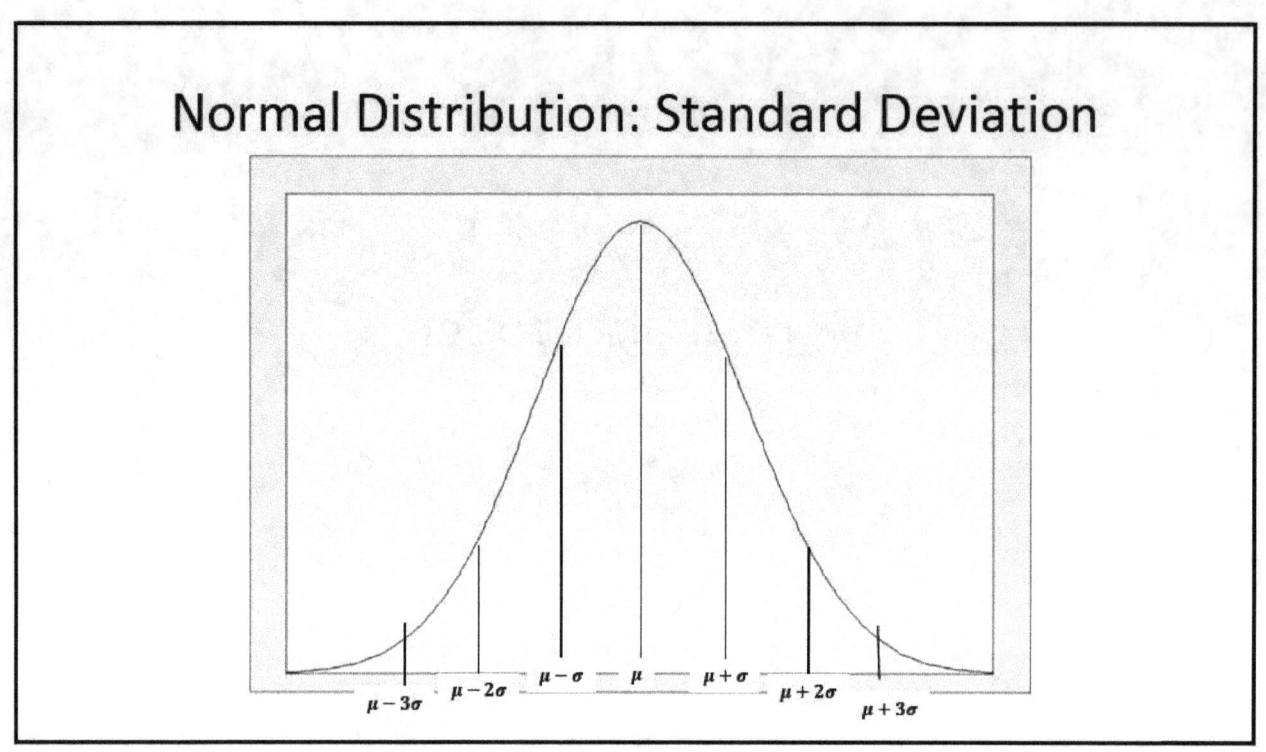

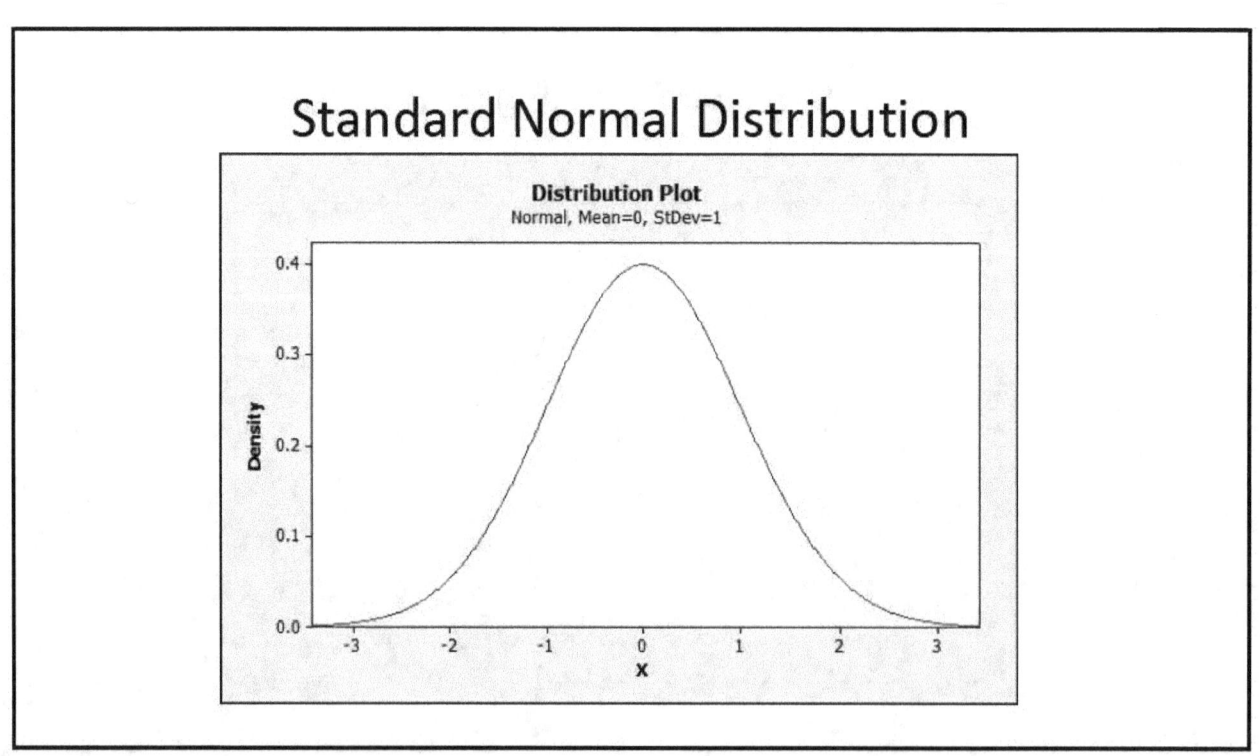

Areas Under the Curve

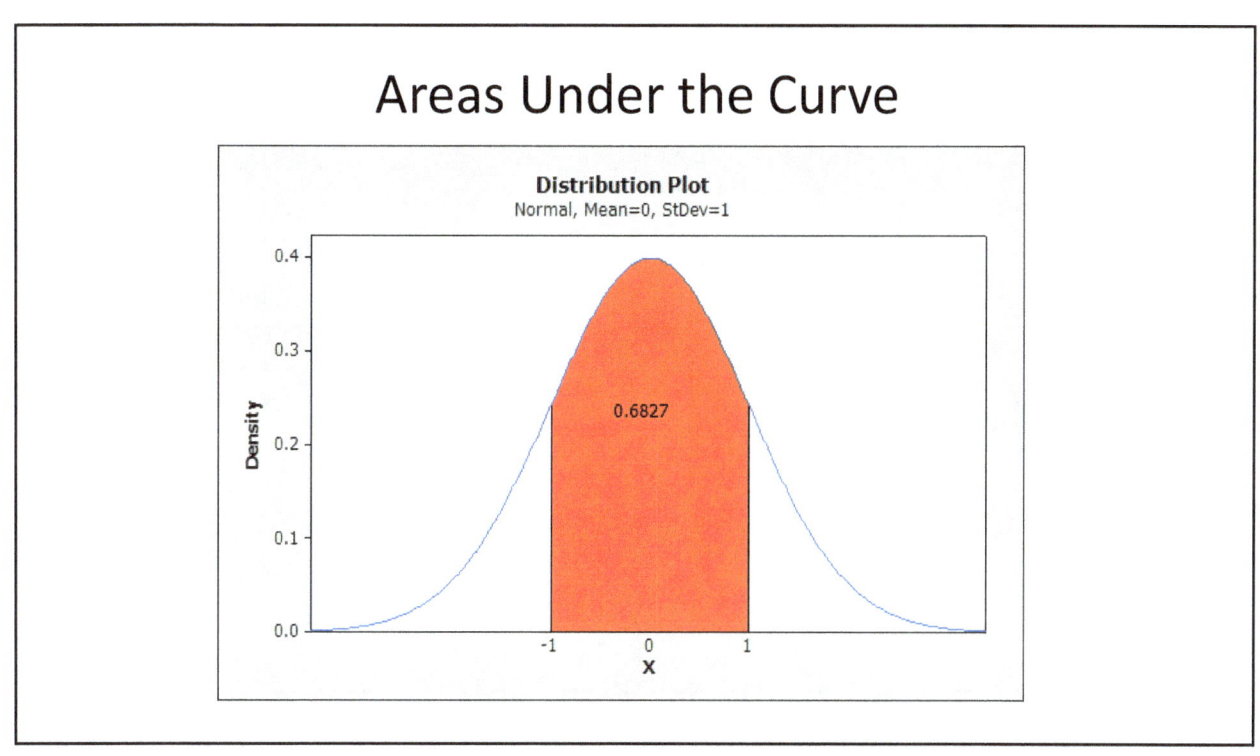

Areas Under the Curve

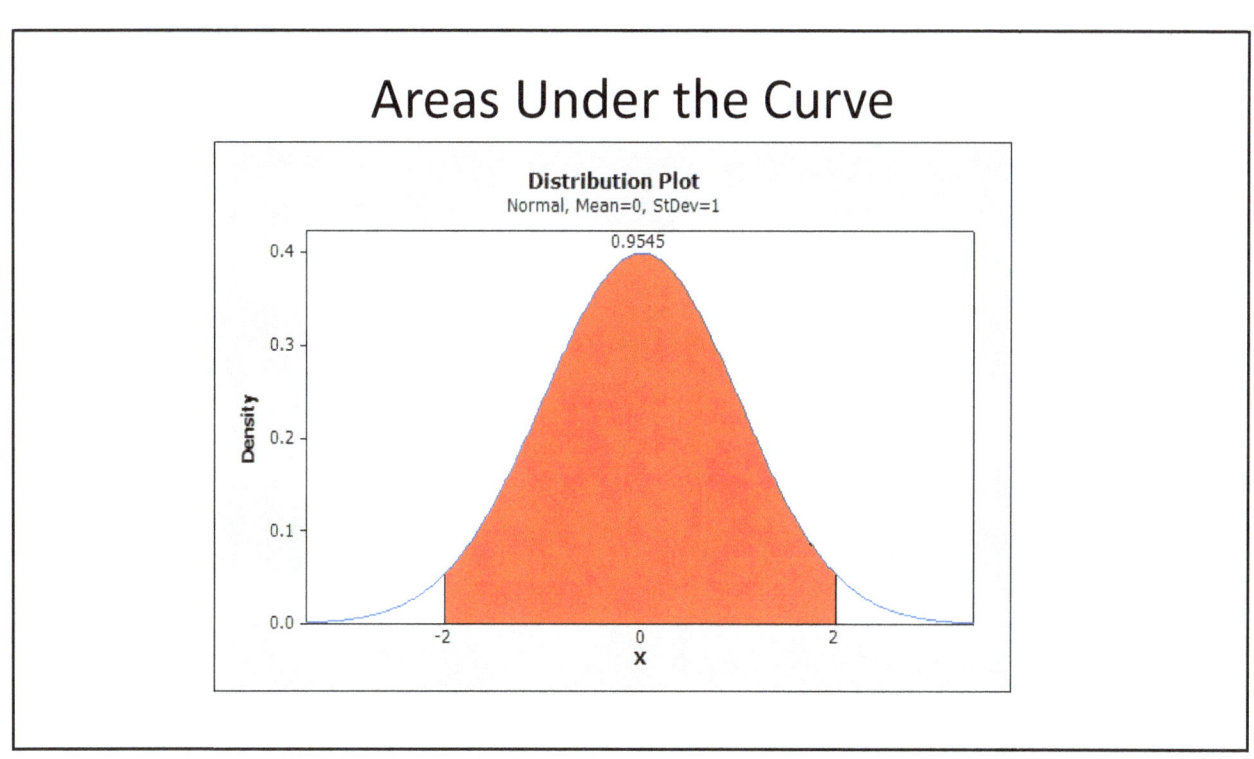

Areas Under the Curve

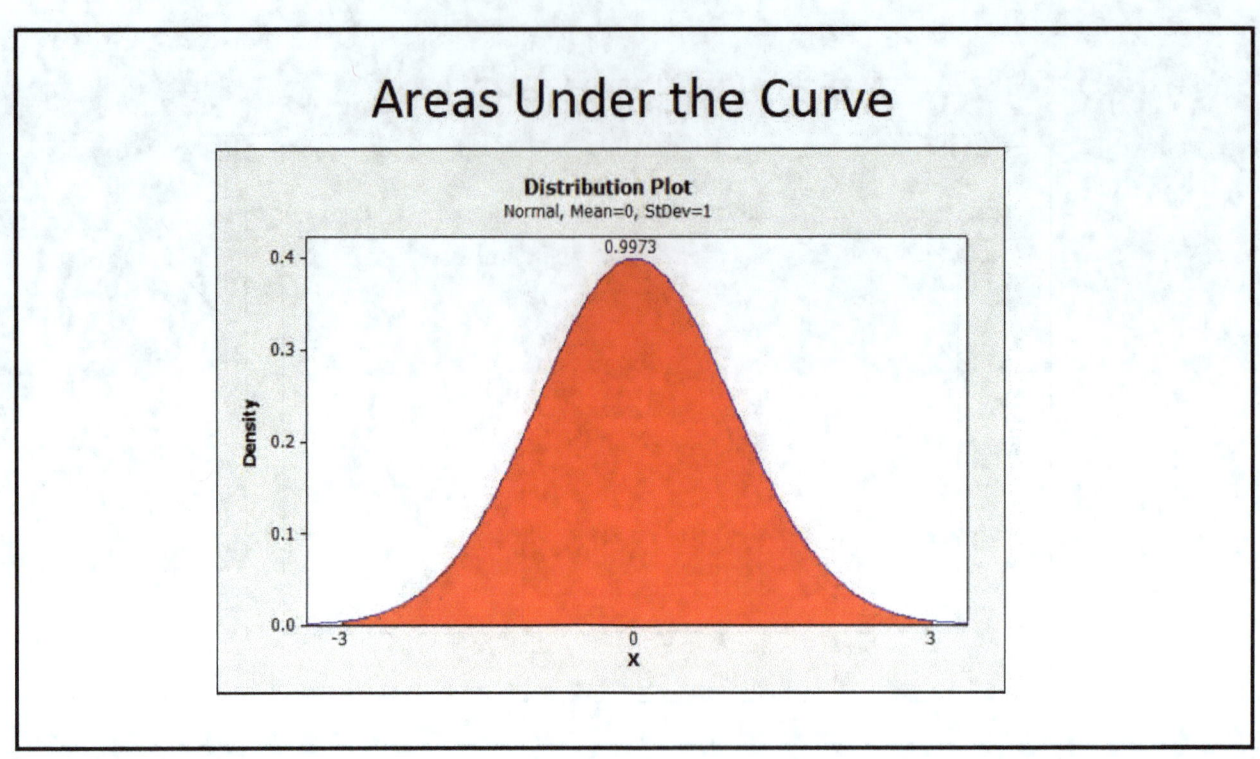

Areas Under the Curve

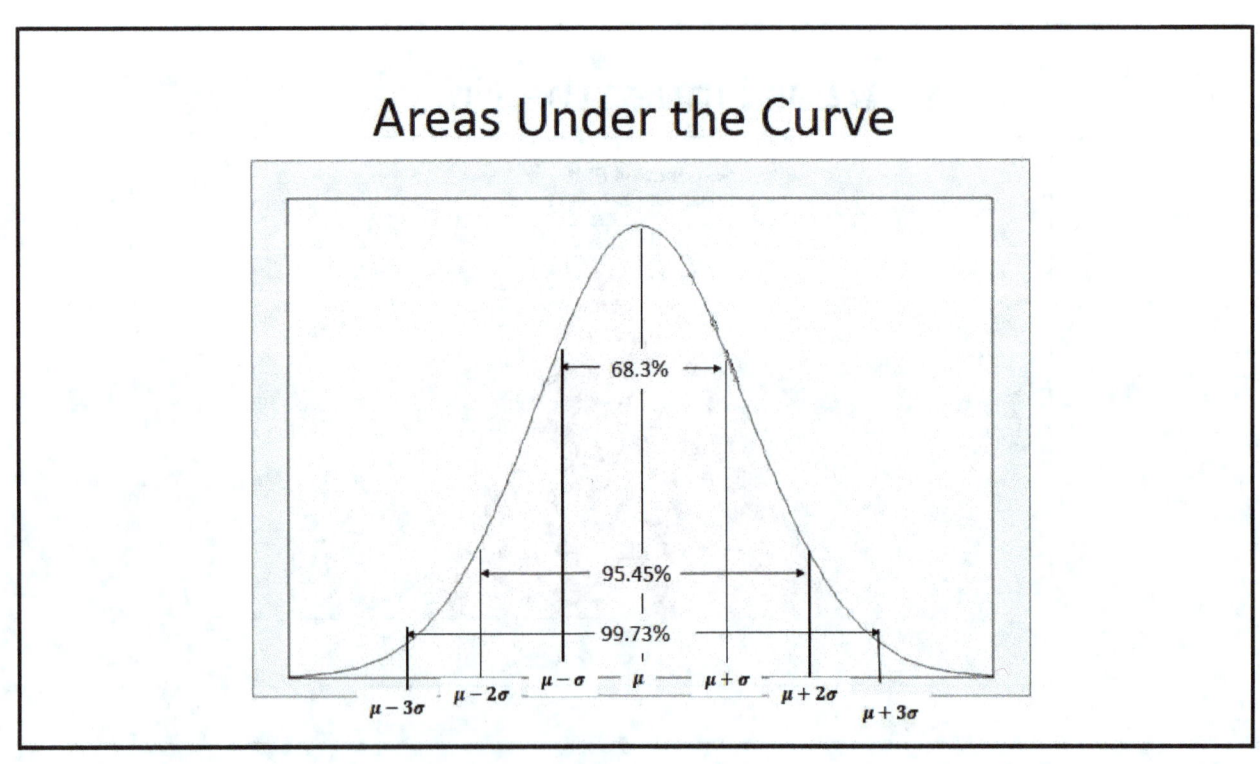

Z Score & Standard Normal

A **Z score** gives us the relative position of a data value on the normal distribution curve.

Given a data value x from a normal distribution with mean μ and standard deviation σ,

$$z = \frac{(x - \mu)}{\sigma}$$

Areas Under the Standard Normal Curve

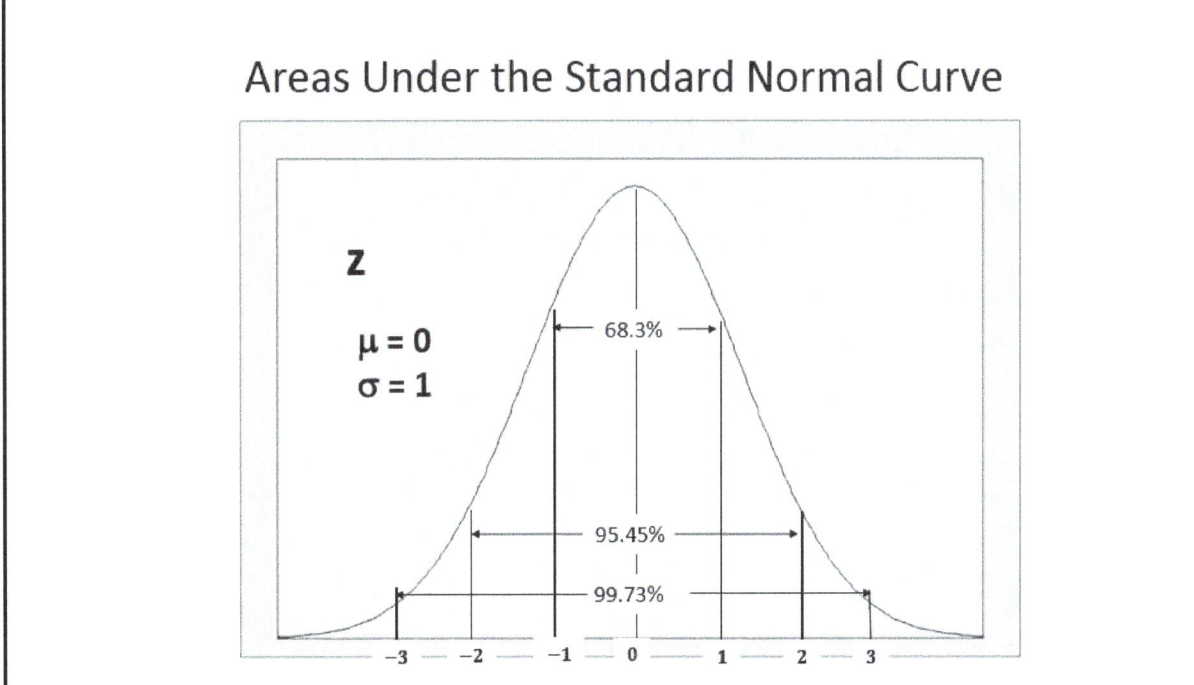

How Unusual?

Rule of thumb:

A **Z score** less than -2 or greater than +2 is considered unusual

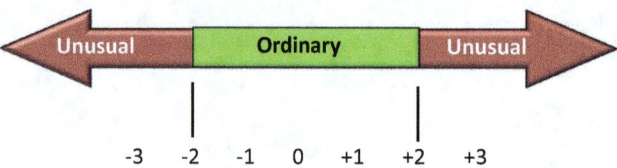

Class Exercise: What's Your Z Score?

Men's heights follow a normal distribution with mean = 69.0 in and standard deviation = 2.8 in.

Women's heights follow a normal distribution with mean = 63.6 in and standard deviation = 2.5 in.

$$z = \frac{x - \mu}{\sigma}$$

Confidence Intervals

How Good is Our Estimate?

- We can estimate the mean using $\overline{X}$

    - This gives us a single point value, but it is very unlikely that the true mean will exactly equal this one point

- A better strategy is to give a range of values that could cover the mean value with a certain probability

    - Takes the uncertainty due to sampling into account

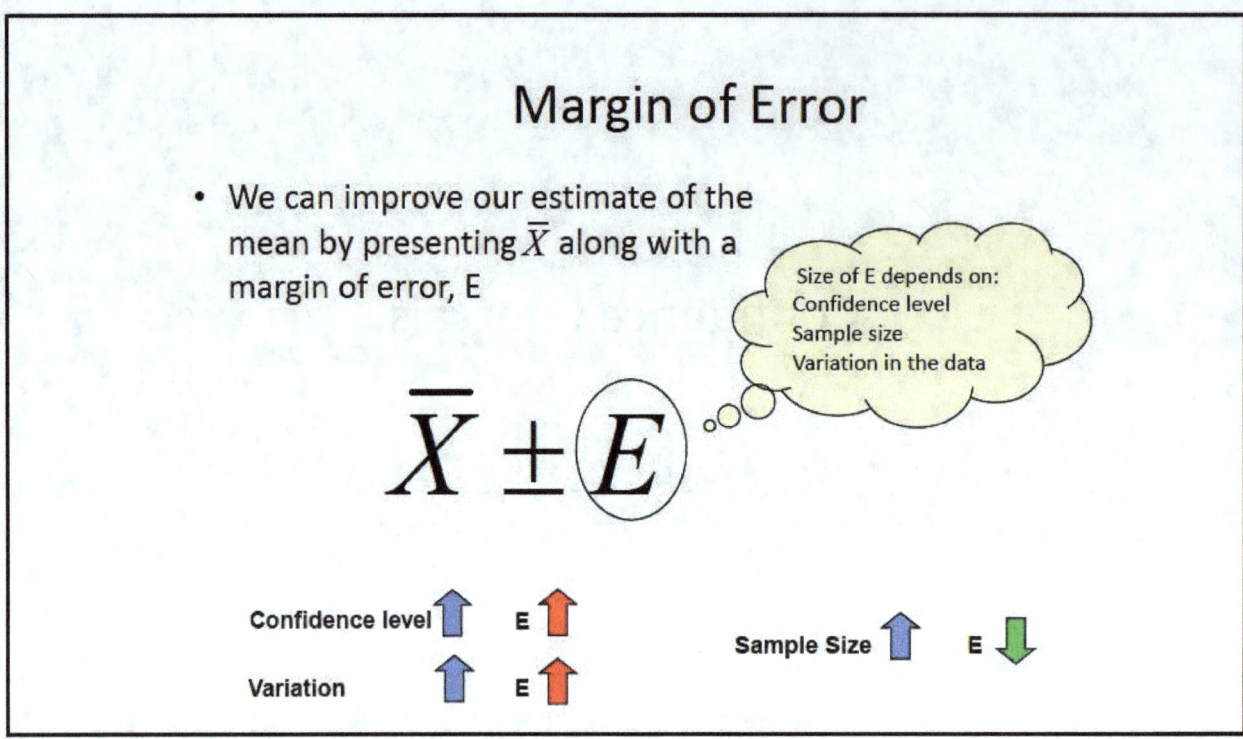

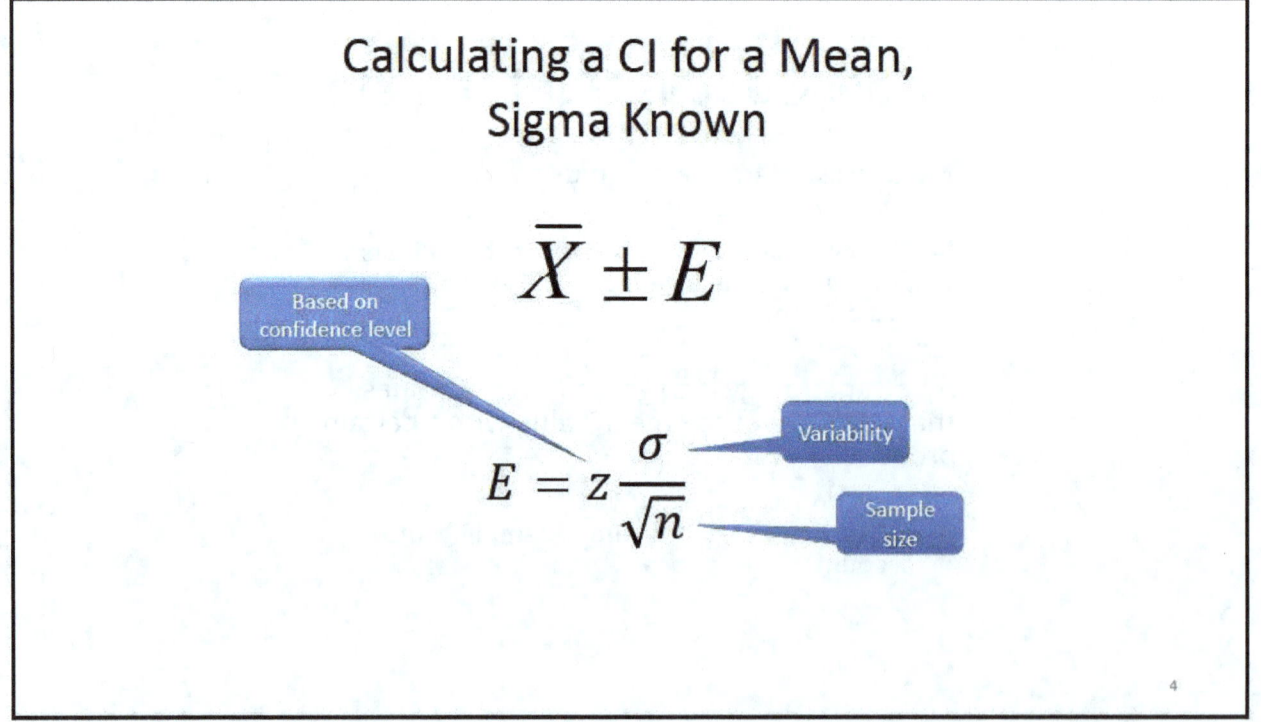

"Famous" Z Values

- For a 90% two-sided CI, use Z = 1.645

- For a 95% two-sided CI, use Z = 1.96

- For a 99% two-sided CI, use Z = 2.575

Confidence Levels

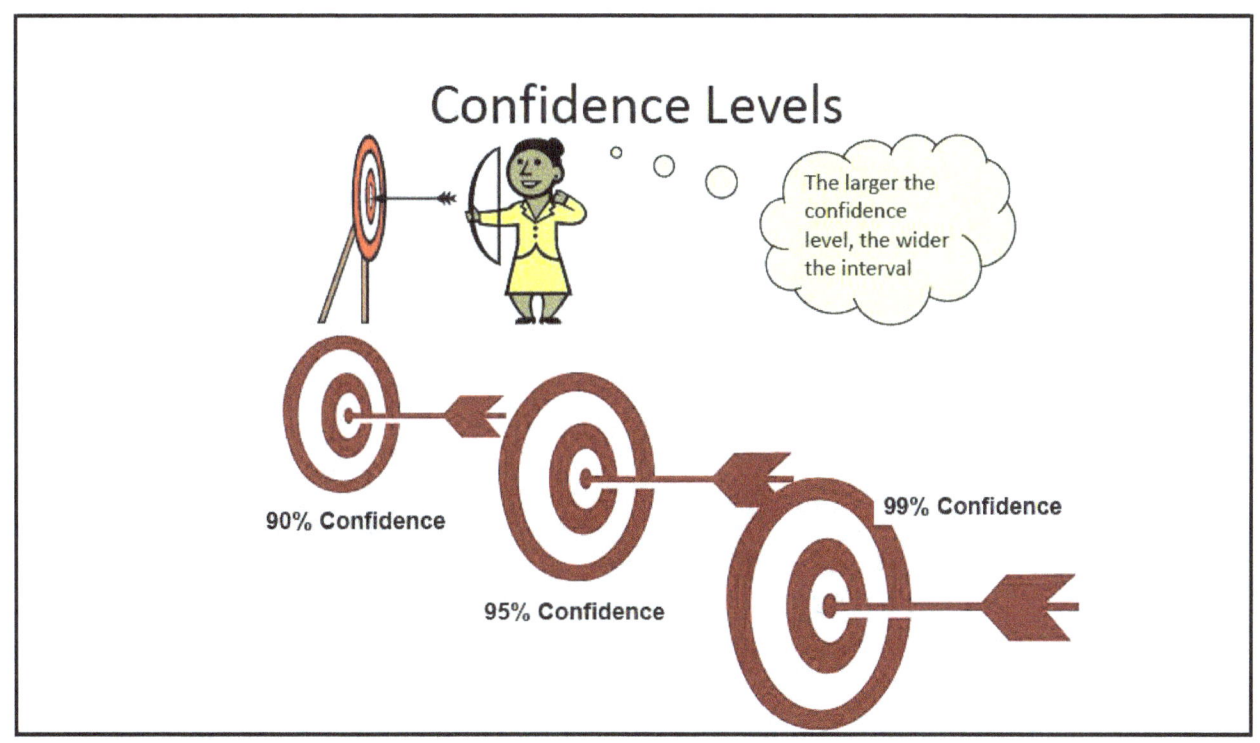

The larger the confidence level, the wider the interval

90% Confidence

95% Confidence

99% Confidence

Calculating a CI for a Mean, Sigma Known

$$\bar{X} \pm E$$

$$E = z \frac{\sigma}{\sqrt{n}}$$

- Based on confidence level
- Variability
- Sample size

Given: $\bar{X} = 24.2$
$\sigma = 10.0$
$n = 30$

1. Construct a 95 % CI about μ
2. Construct a 99 % CI about μ

It's All in the Coverage

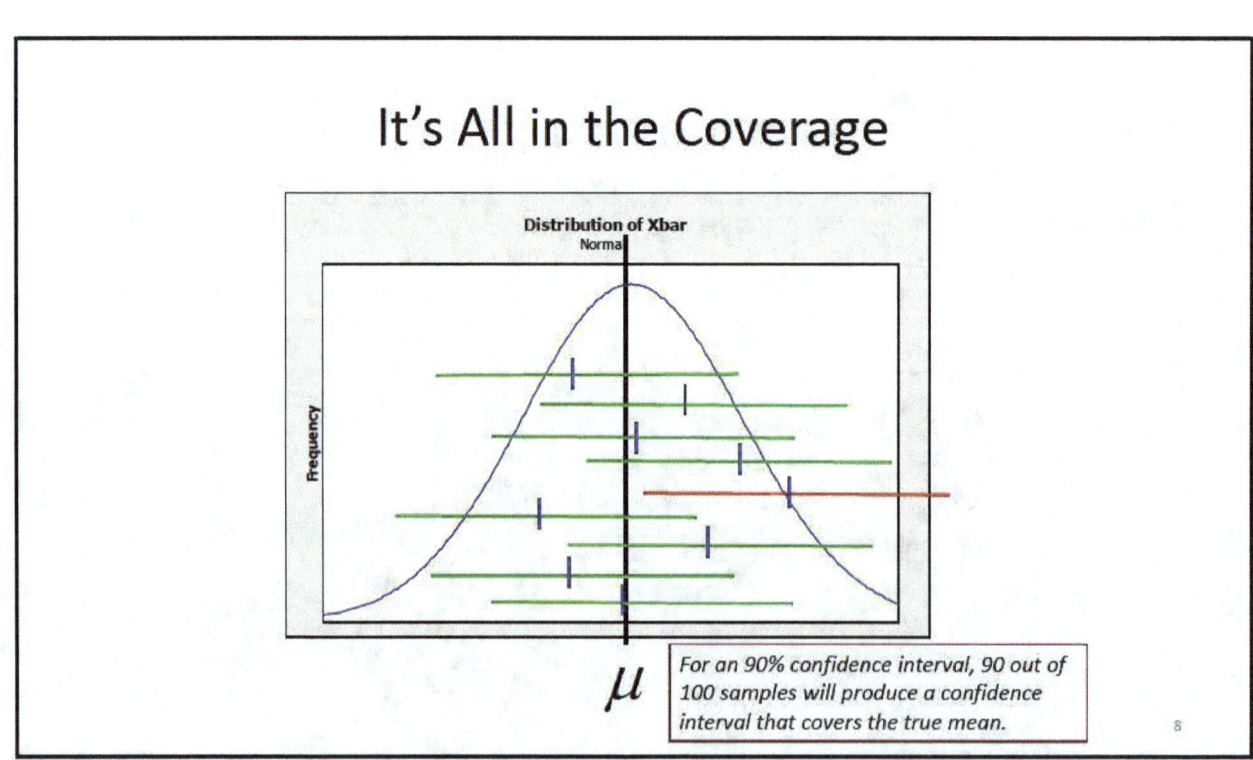

For an 90% confidence interval, 90 out of 100 samples will produce a confidence interval that covers the true mean.

Interpreting a Confidence Interval

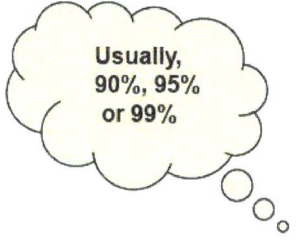
Usually, 90%, 95% or 99%

The mean NEVER falls.

"We are ___ % confident that the true mean is covered by the interval ____, ____."

Process Capability

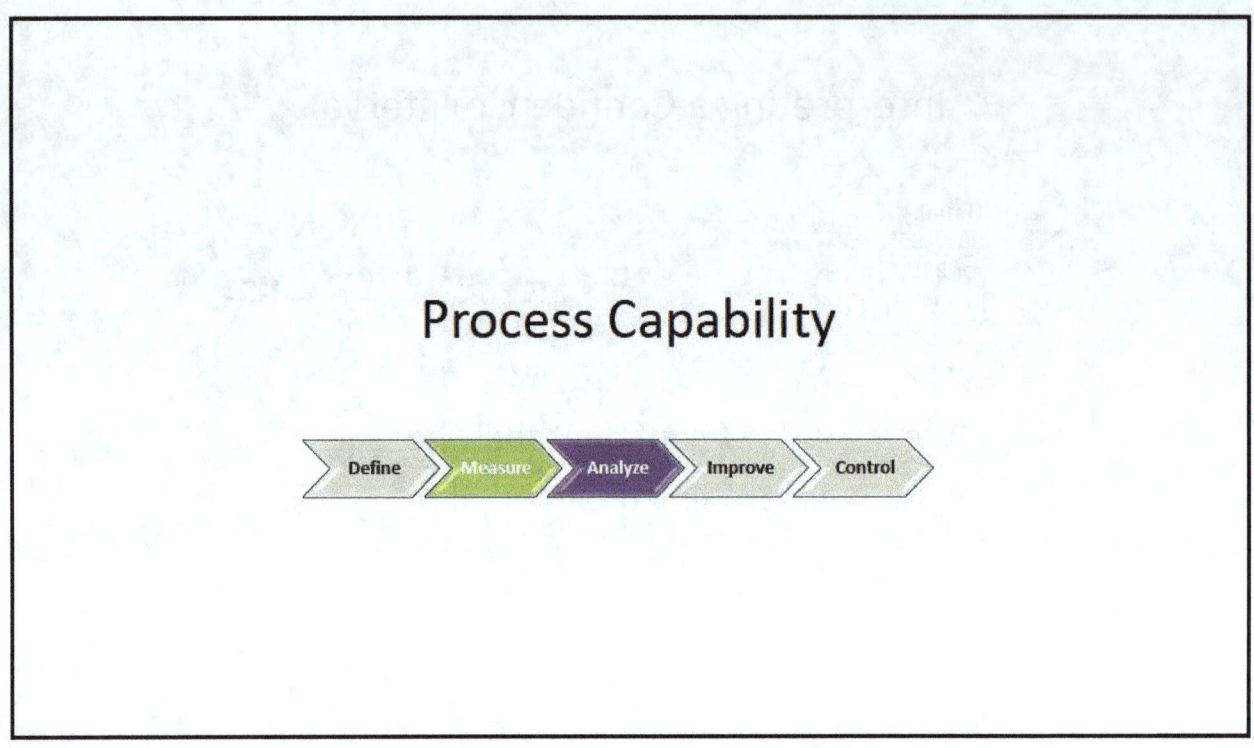

Capability Indices

- **Capability** is a measure of the inherent uniformity of the process

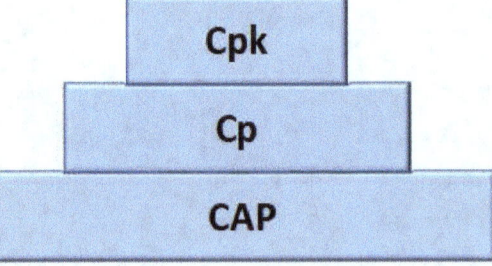

Capability Requirements

- The process must be predictable (in-control)

- The process metric must be normally distributed

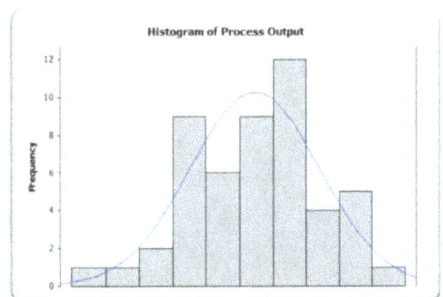

CAP

- **CAP** is also known as the **natural tolerance**
- CAP is a measure of the total spread of the process

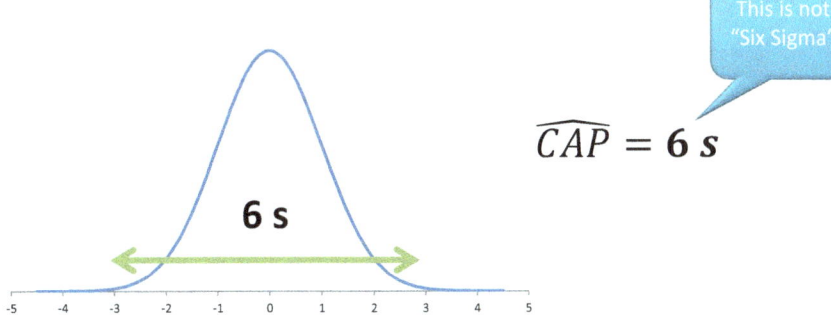

$$\widehat{CAP} = 6\,s$$

This is not "Six Sigma"

Cp

Cp compares the specification width to the natural tolerance of the process

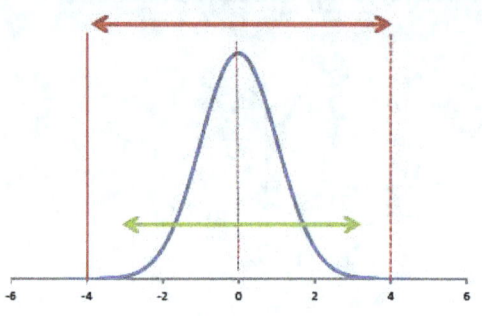

$$\hat{C}_p = \frac{(USL - LSL)}{6s}$$

Cp Examples

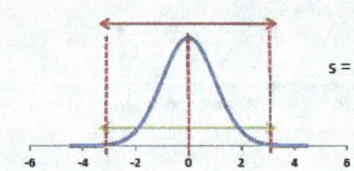
s = 1
$$\hat{C}_p = \frac{(USL - LSL)}{6s} = 1$$

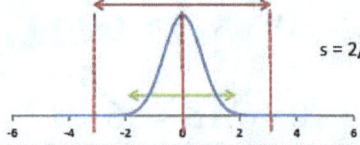
s = 2/3
$$\hat{C}_p = 1.5$$

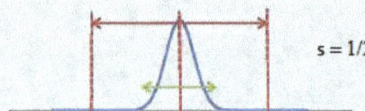
s = 1/2
$$\hat{C}_p = 2.0$$

Cp

Compare the Cp values of these processes

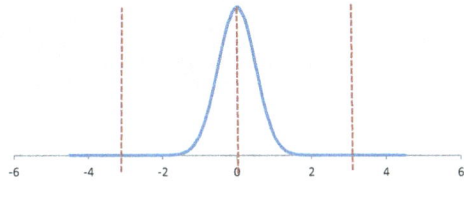

$\hat{C}_p = 2.0$

$\hat{C}_p = 2.0$

Cpk

- **Cpk** compares both the spread and the location of the process to the specifications

$$\hat{C}_{pl} = \frac{(\mu - LSL)}{3s} \qquad \hat{C}_{pu} = \frac{(USL - \mu)}{3s}$$

$$\hat{C}_{pk} = min(\hat{C}_{pl}, \hat{C}_{pu})$$

Cpk

For a centered process, Cp = Cpk

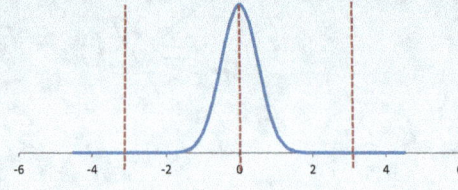

$\hat{C}_{pk} = 2.0$

$\hat{C}_p = 2.0$

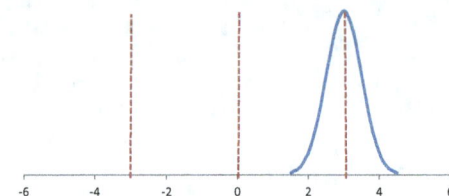

$\hat{C}_{pk} = 0$

$\hat{C}_p = 2.0$

Capability and Sigma Level

Sigma	Cp	DPMO	Yield (%)
6.0	2.00	0.0	100.0000
5.0	1.67	0.6	99.9999
4.0	1.33	63.3	99.9937
3.0	1.00	2,700	99.7300
2.0	0.67	45,500	95.4500
1.0	0.33	317,311	68.2689

Does not include the 1.5 sigma shift

Capability Examples

Shuttle wait times

Based on customer feedback, Mirasol has set the upper limit for wait time at the ferry dock to 15 minutes, with a target of 9 minutes and a lower limit of 3 minutes. Waiting a few minutes for the shuttle gives arriving guests a chance to stop at the water sports kiosk adjacent to the dock. The kiosk has information on dolphin cruise schedules, deep sea fishing excursions and rentals on jet skis, kite boards, kayaks and boats.

- Draw in the customer requirements: target and upper and lower limits
- Comment on the shape, center and variation of the data. Is it skewed? Centered? What is the spread?
- Assuming stability, can capability be calculated here? If so, estimate CAP, Cp and Cpk.

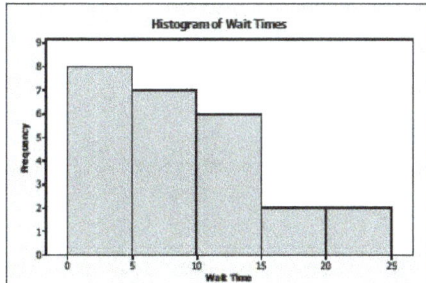

Laundry Facility Water Usage

Water and wastewater disposal comprise half of the total costs of a commercial laundry facility. The measure of efficiency for commercial laundry facilities is measured in gallons of water per pound of fabric. This metric was recorded for 25 consecutive wash cycles, with the results shown below. The service contractor claims that the current wash cycle design should have a target of 4.8 gal/lb with lower limit = 4.5 gal/lb and an upper limit = 5.1 gal/lb.

- Draw in the customer requirements on the histogram: target and upper and lower limits
- Comment on the shape, center and variation of the data. Is it skewed? Centered? What is the spread?
- Assuming stability, can capability be calculated here? If so, estimate CAP, Cp and Cpk.

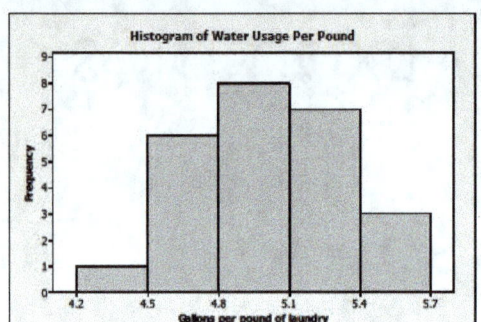

N	25
Sample Mean	5.0
Sample Standard Deviation	0.3
Minimum	4.3
Maximum	5.6

Target	4.8
Lower specification	4.5
Upper specification	5.1

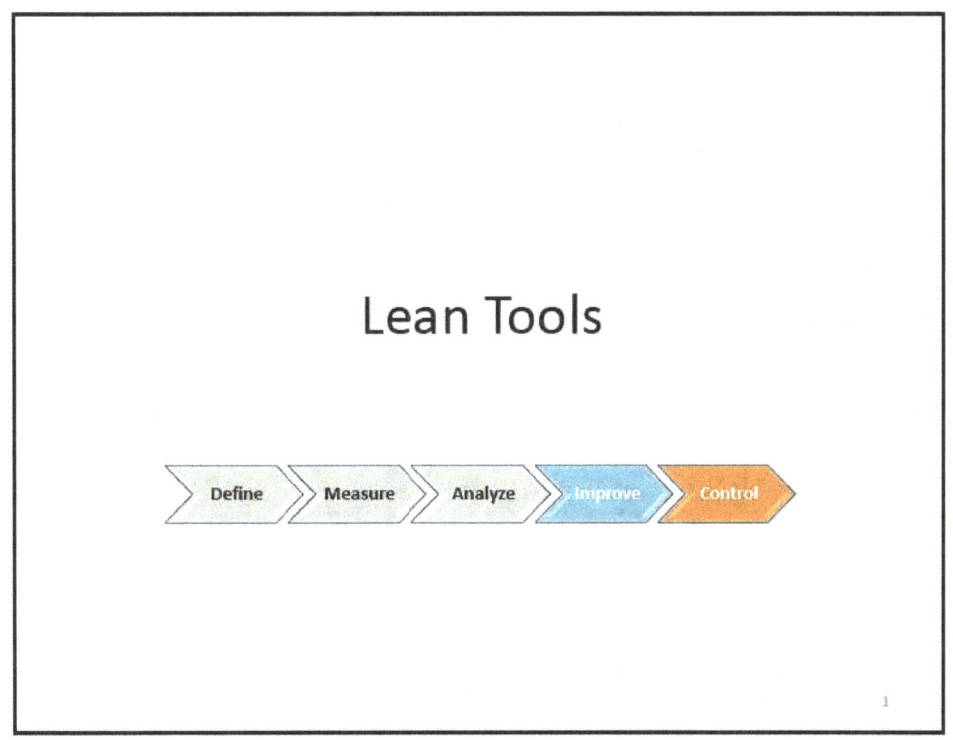

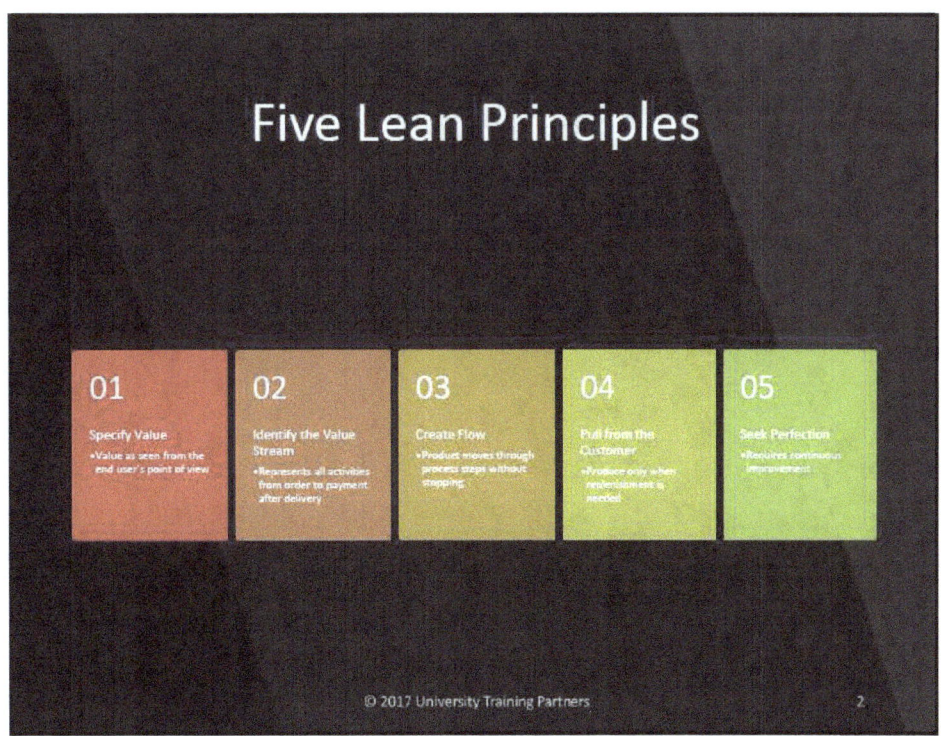

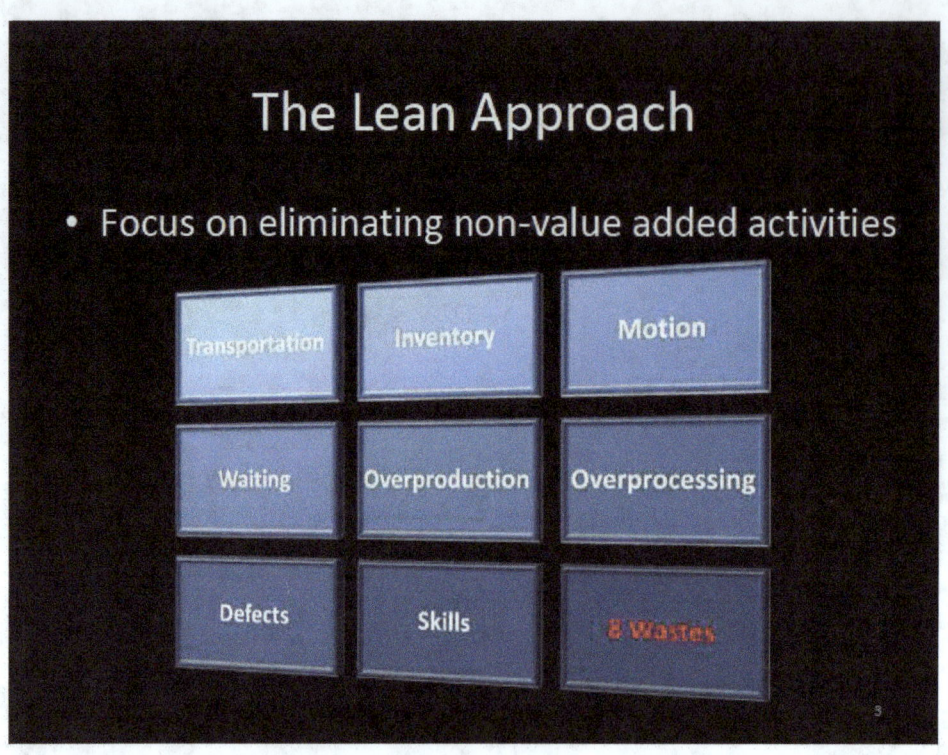

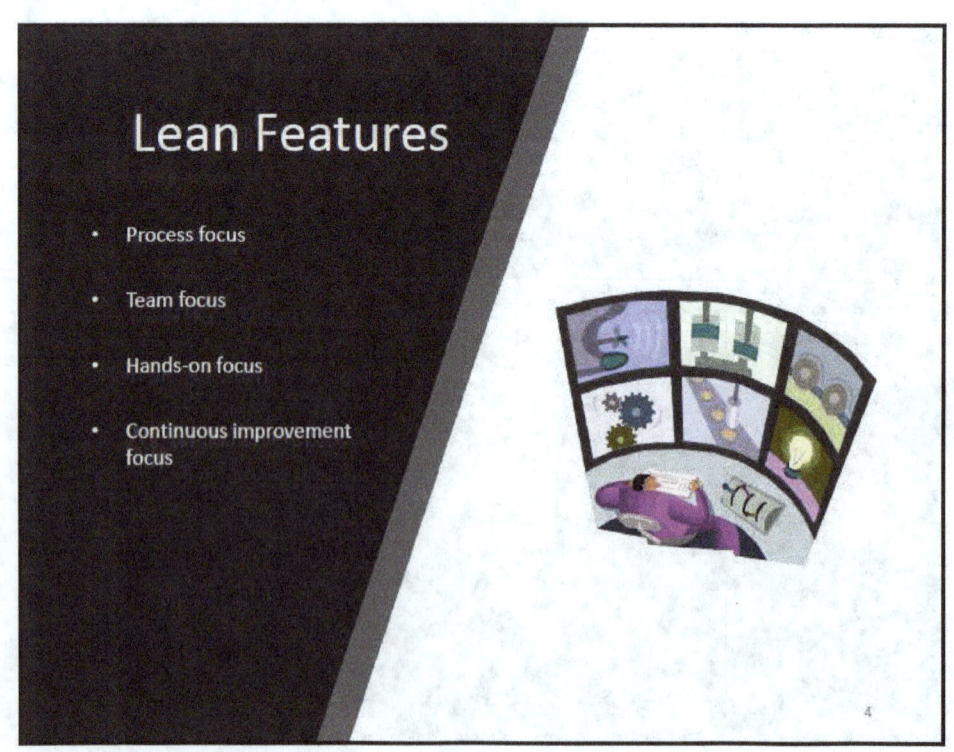

Lean Results in...

- Faster cycle times
- Faster change over times
- Little to no inventory
- Just-in-time methods
- Pull systems
- Continuous flow
- Production leveling
- High quality levels
- Higher employee morale
- **Reduced costs**
- **Higher customer satisfaction**

The Lean Enterprise is known for speed, flow and quality.

Visual Management

Visual Management makes the current status of key inputs, outputs and processes apparent at a glance.

Inexpensive

Simple

Unambiguous

Immediate

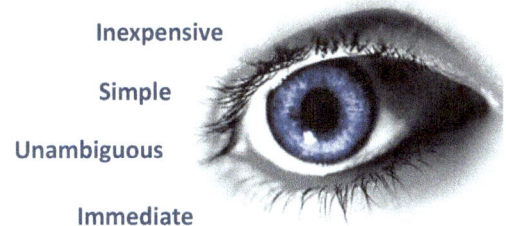

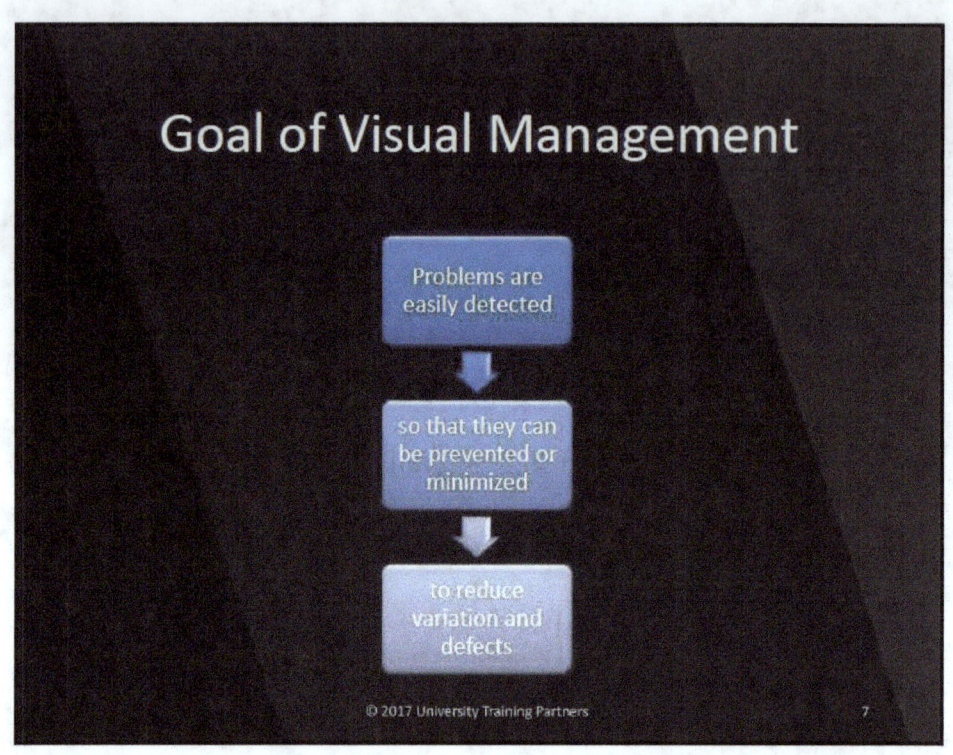
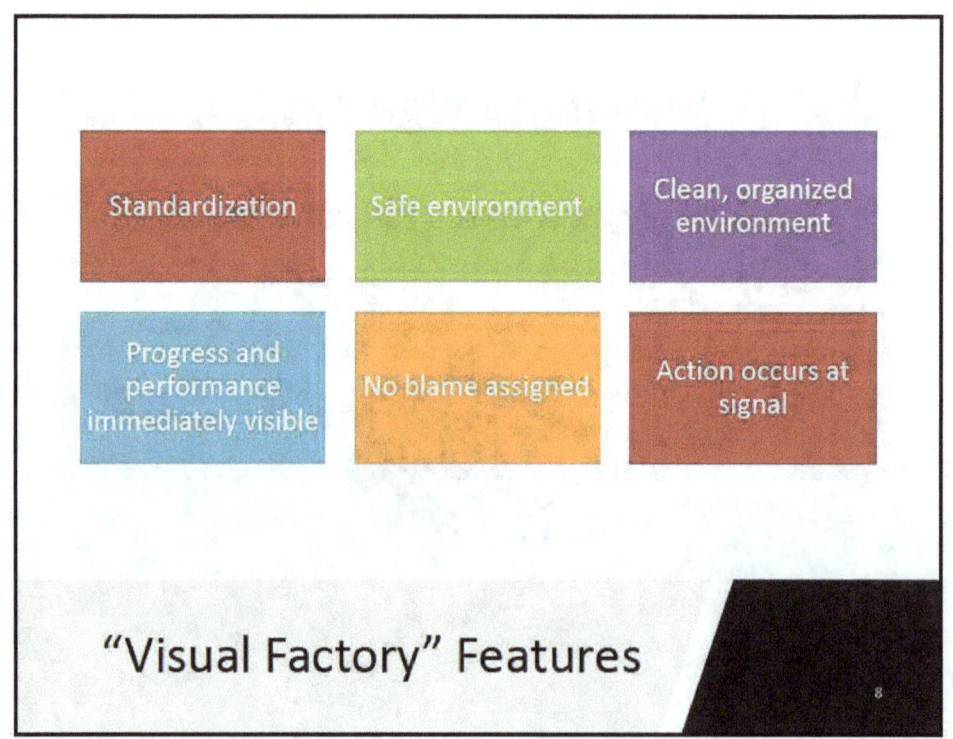

Visual Management Tools

- Color coding – floor tape, racks, bins
- Machine indicator lights (Andon)
- Operator call lights
- Work instructions
- Maintenance charts
- SPC Charts
- Kanban bins
- Display boards

Color Coding

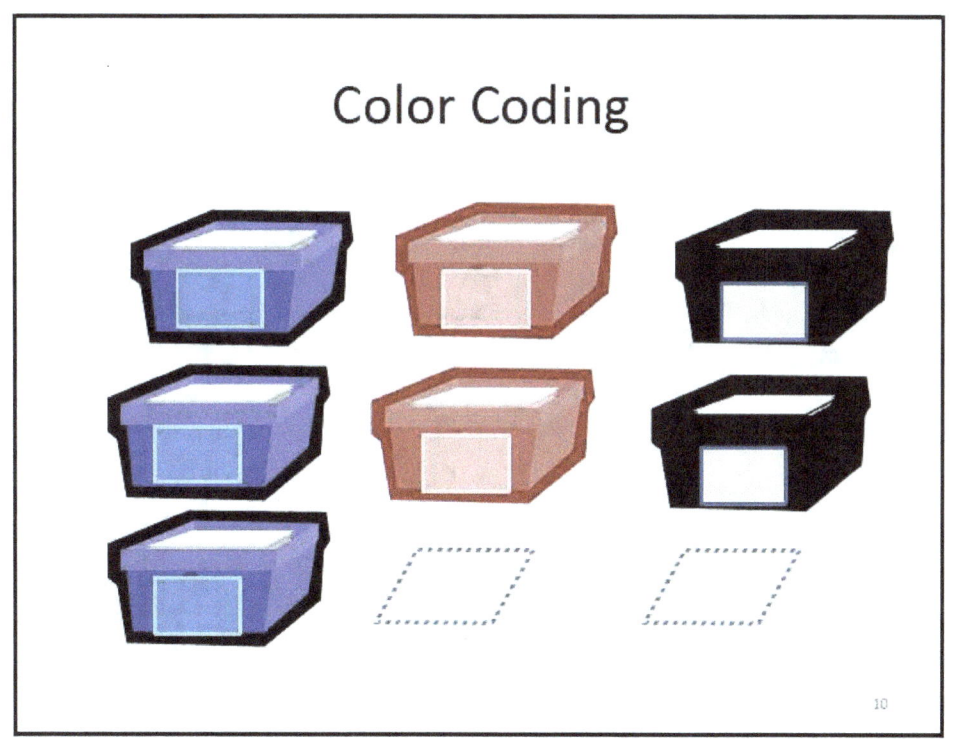

Eight Wastes

- Focus on eliminating non-value added activities

Transportation	Inventory	Motion
Waiting	Overproduction	Over-processing
Defects	Skills	**8 Wastes**

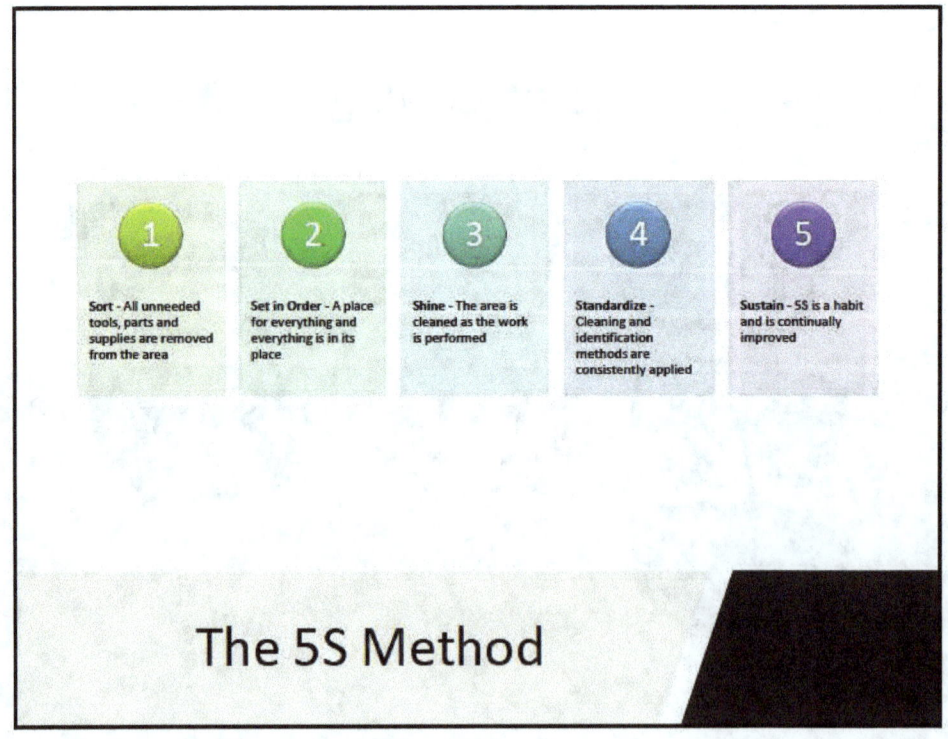

The 5S Method

1. **Sort** - All unneeded tools, parts and supplies are removed from the area
2. **Set in Order** - A place for everything and everything is in its place
3. **Shine** - The area is cleaned as the work is performed
4. **Standardize** - Cleaning and identification methods are consistently applied
5. **Sustain** - 5S is a habit and is continually improved

Advantages of 5S

- Productivity (Speed)
- Reduced variation
- Problem identification

Sort
Set in order
Shine
Standardize
Sustain

Mistake Proofing

- Also known as **poka yoke**
- Especially useful in preventing defects when they are rare
- Effective in eliminating random, non-systematic errors

Types of Mistake Proofing

- Control or Warning
  - Machine or process shuts down or signals when an error occurs
  - Lights or sounds alert operator
  - Prevents defects from moving on to the next stage
- Prevention
  - Design does not allow defects from occurring in the first place
  - 100% elimination of defect type

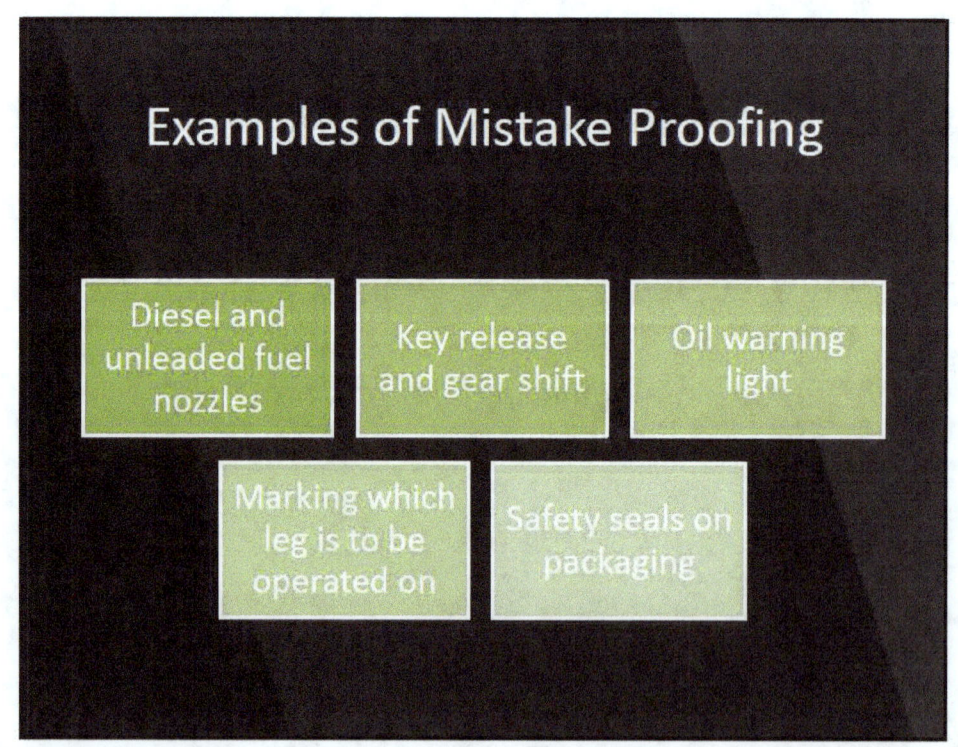

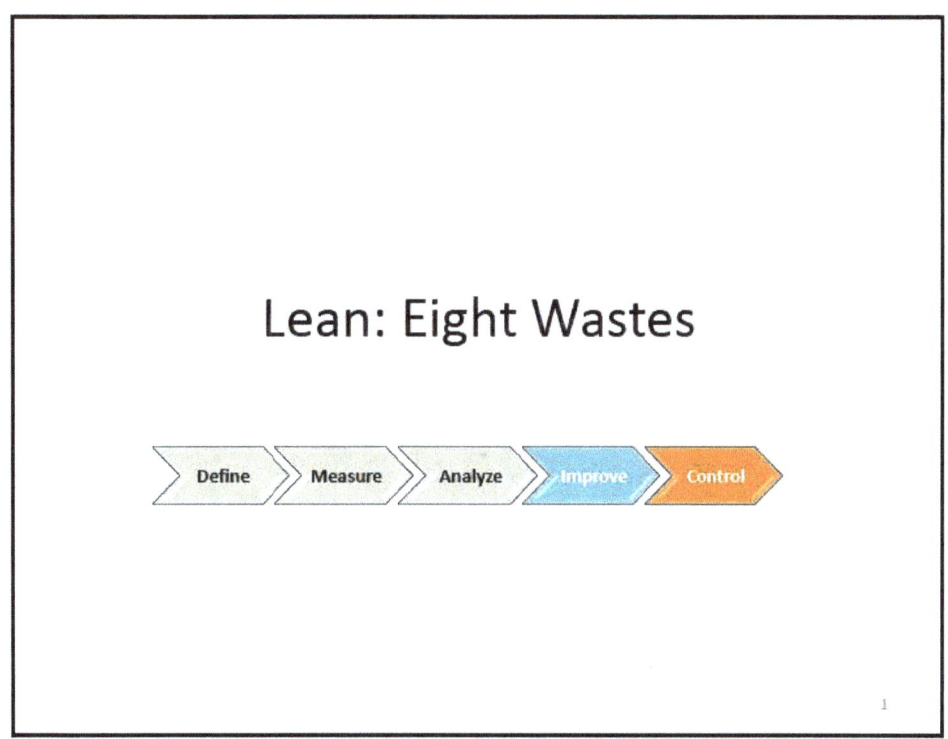

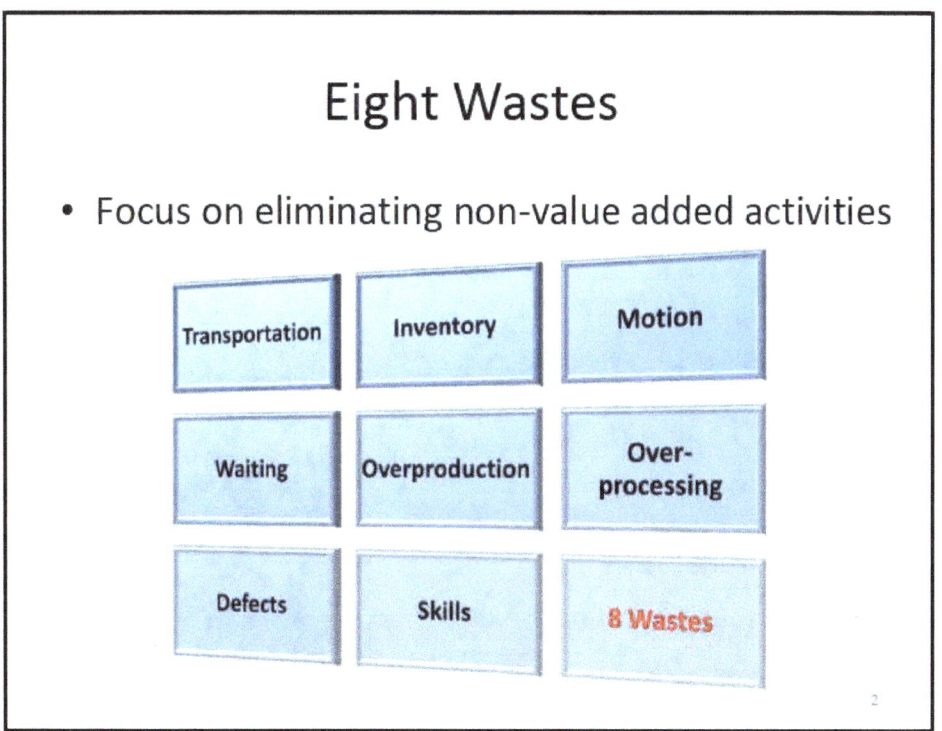

Transportation

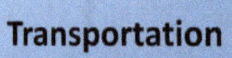

Definition:
Movement of things:
- Paperwork
- Material
- Electronic information

Risks:
Damage or loss
Delay in availability

Inventory

Definition:
Excess:
- Paperwork
- Material
- Supplies
- Equipment

Risks:
Reduced cash flow
Lost productivity due to searching
Excess space
Damage
Obsolescence

Signs:
Stockpiles of supplier, forms, materials
Disorganized storage areas

Motion

Definition:
Movement of people

Signs:
Hand carrying
Functional layout
Traveling to shared equipment

Risks:
Reduced capacity
Injury

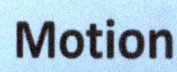

Waiting

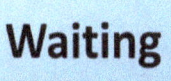

Definition:
People waiting for people, information, product or machines

Information, product or machine waiting for people

Signs:
Idle people
Idle equipment

Risks:
Long lead times
Long work hours
Paid overtime
Capital expenditures for equipment

Overproduction

Definition:
Producing too much too soon

Risks:
Long lead times
Increased complexity

Signs:
Build up of work-in-process (WIP)

Build up of queues of material or people

Over-processing

Definition:
Doing more than the customer is willing to pay for

Risks:
Long lead times
Low productivity
Frustrated workforce

Signs:
Inspections & audits
Redundant tasks
Too many handoffs

Defects

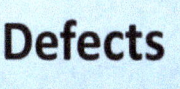

Definition:
Internal or external suppliers provide incorrect, incomplete or late information or material

Signs:
Correcting material or information that has been supplied
Adding missing information
Clarifying information

Risks:
Rework
Scrap
Dissatisfied customers
Long lead times
Warranty costs

Skills

Definition:
Not using people's knowledge, skills, aptitude or creativity

Risks:
Frustrated workforce
Absenteeism
Turnover

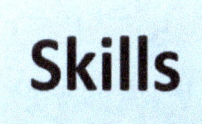

Signs:
Excessive reviews or approvals
Specialized workers
Processes designed by managers
Excessive hand-offs

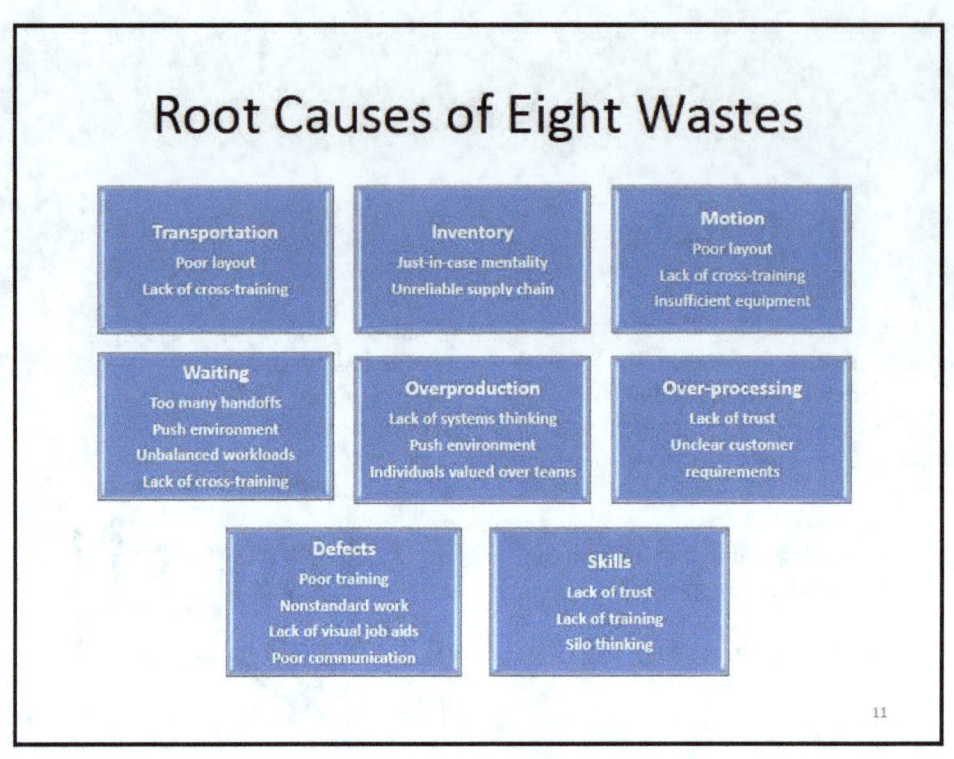

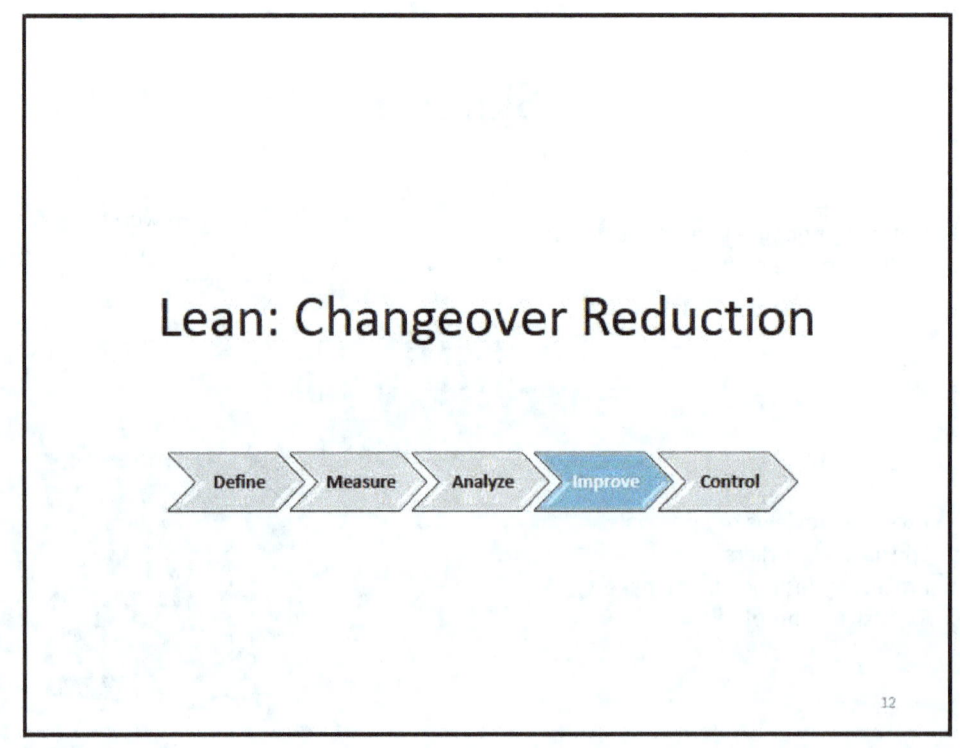

Changeover Reduction

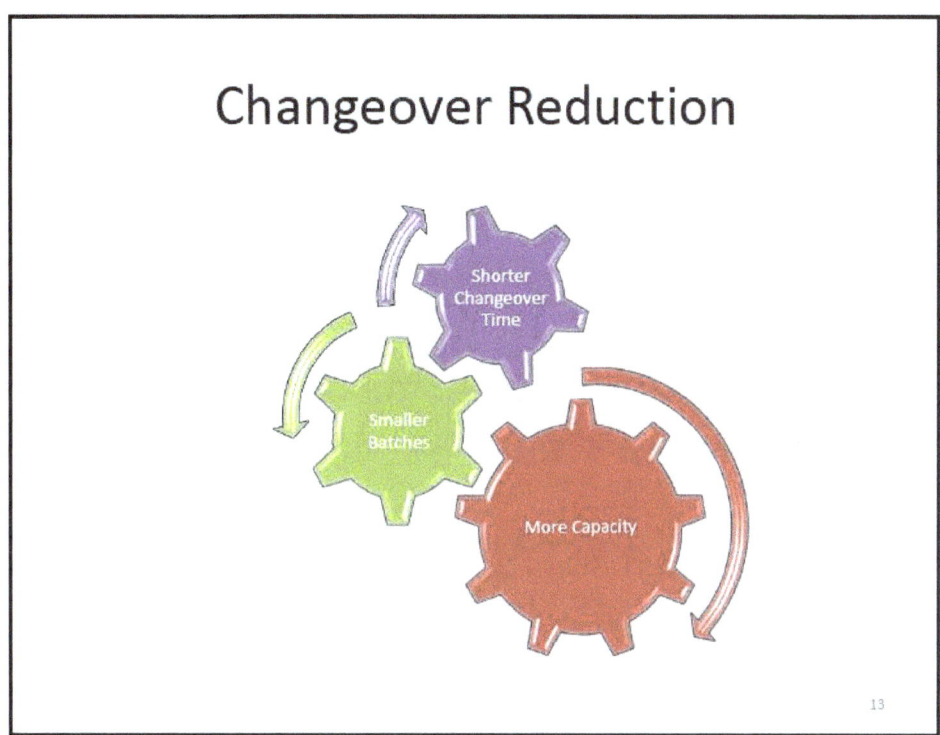

Changeover Time

- **Changeover time** is the total time from the **last unit of production** to the **first unit of good production at full speed**

SMED

- Dr. Shigeo Shingo developed the **"Single Minute Exchange of Dies"** (SMED) concept

- Changeover was reduced from hours to under ten minutes ("single minute")

- Convert internal work to external work

SMED Approach

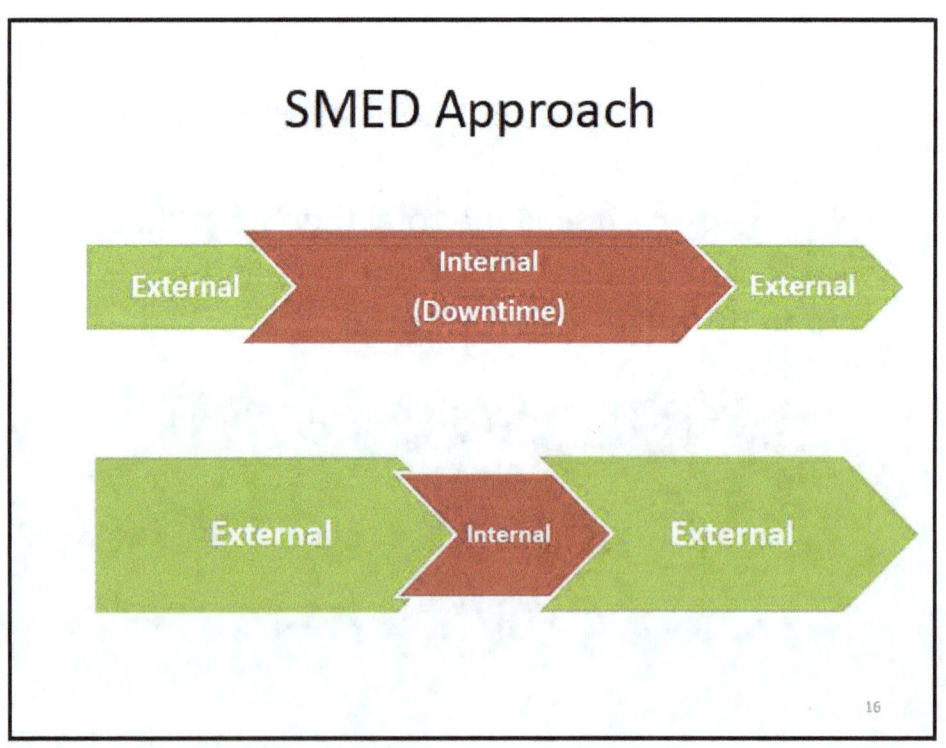

Formula 1 Pit Crew

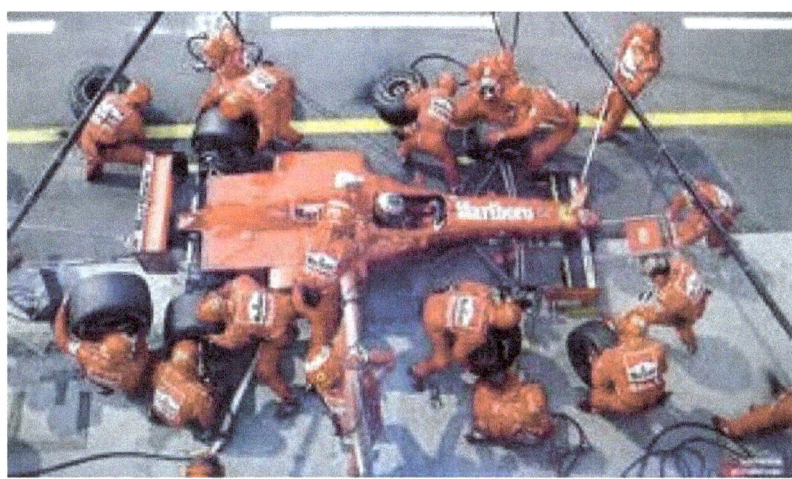

SMED Methods

- Staging carts
- Transport fixtures
- Alignment guides
- Quick fasteners
- Interchangeable parts
- Work breakdown and allocation
- Practice, practice, practice

Redesign, Reengineer, Retool

Lean: Value Stream Mapping

Value Stream Maps

- Detailed process map that captures all key flows (information, work, materials) and process metrics
- Used to identify and quantify waste
- Current state maps help identify project opportunities
- Future state maps help create a blueprint for improvement

Creating a Current State VSM

1. Choose a product, service or family to map
2. Draw the process flow
3. Add the material flow
4. Add information flow
5. Collect process data and add to chart
6. Add process and lead time data to chart
7. Verify the map, identify waste

Creating a Future State VSM

Now create a map showing the ideal future state, and place what must happen to change to future state in Kaizen Bursts.

This is your plan for the next 6 months to one year.

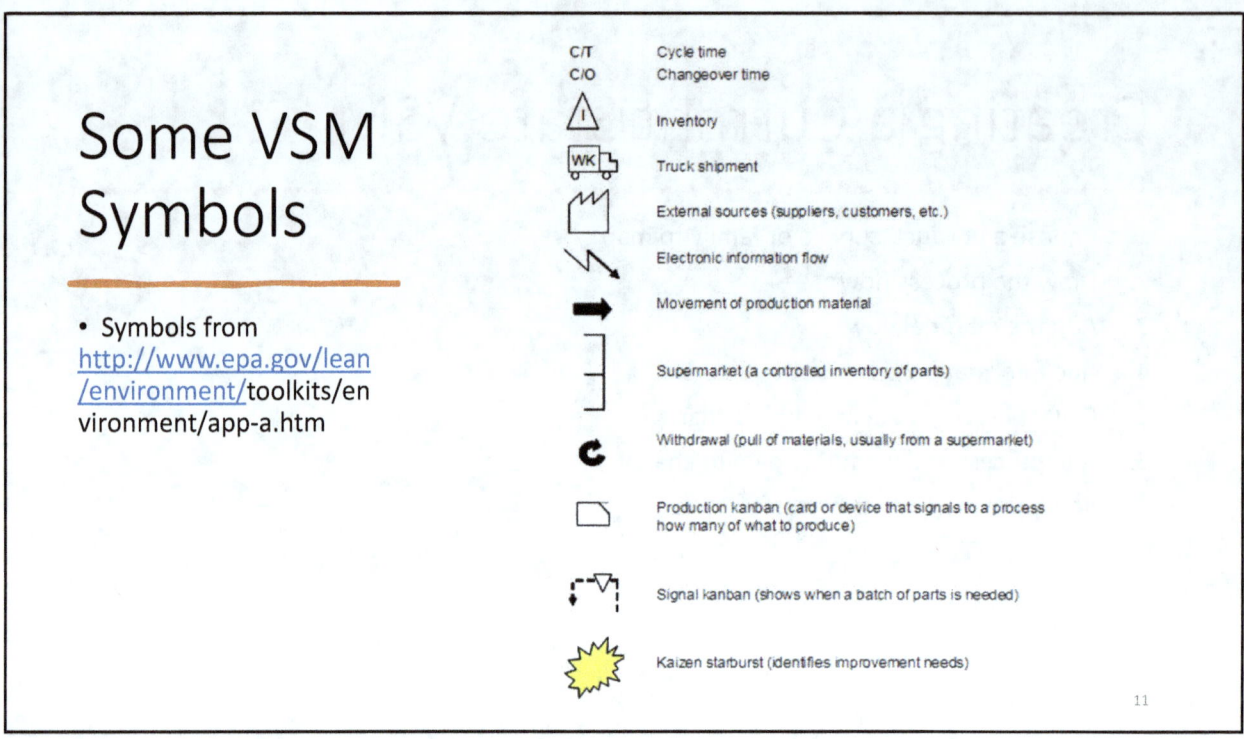

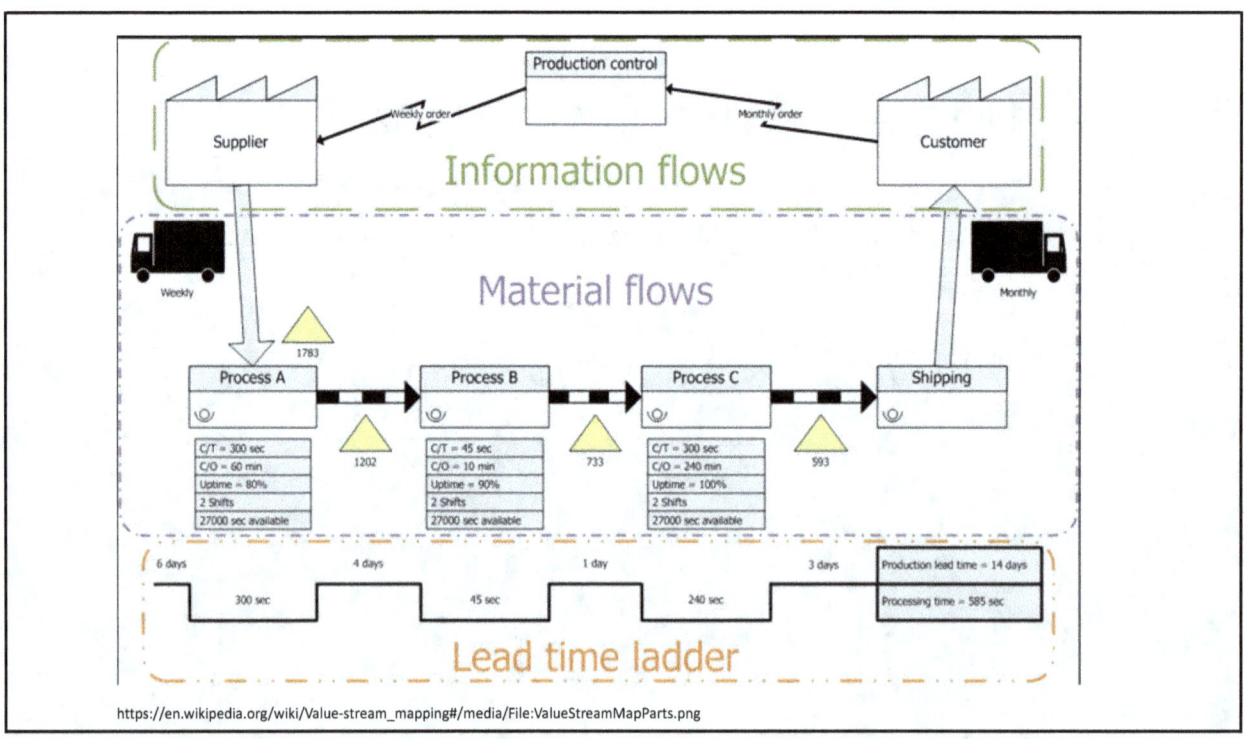

Kaizen Event Planning

Characteristics of a Kaizen Event

- Value Stream Driven
- Total employee involvement
- Cross-functional teamwork
- 100% focus
- Short duration
- Aggressive objectives
- Creativity before capital
- Waste elimination
- Full implementation
- New process training
- Built-in sustainability
- Workforce development

Source: Martin & Osterling, The Kaizen Event Planner, CRC Press, 2007

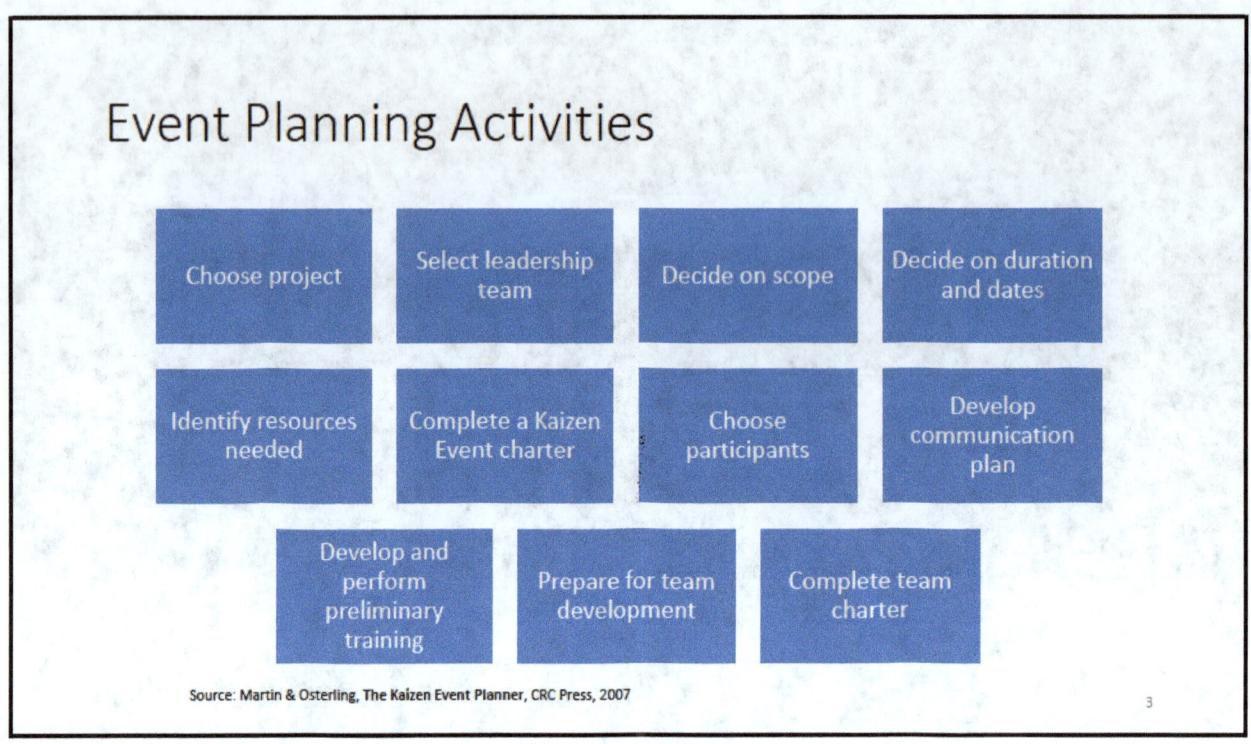

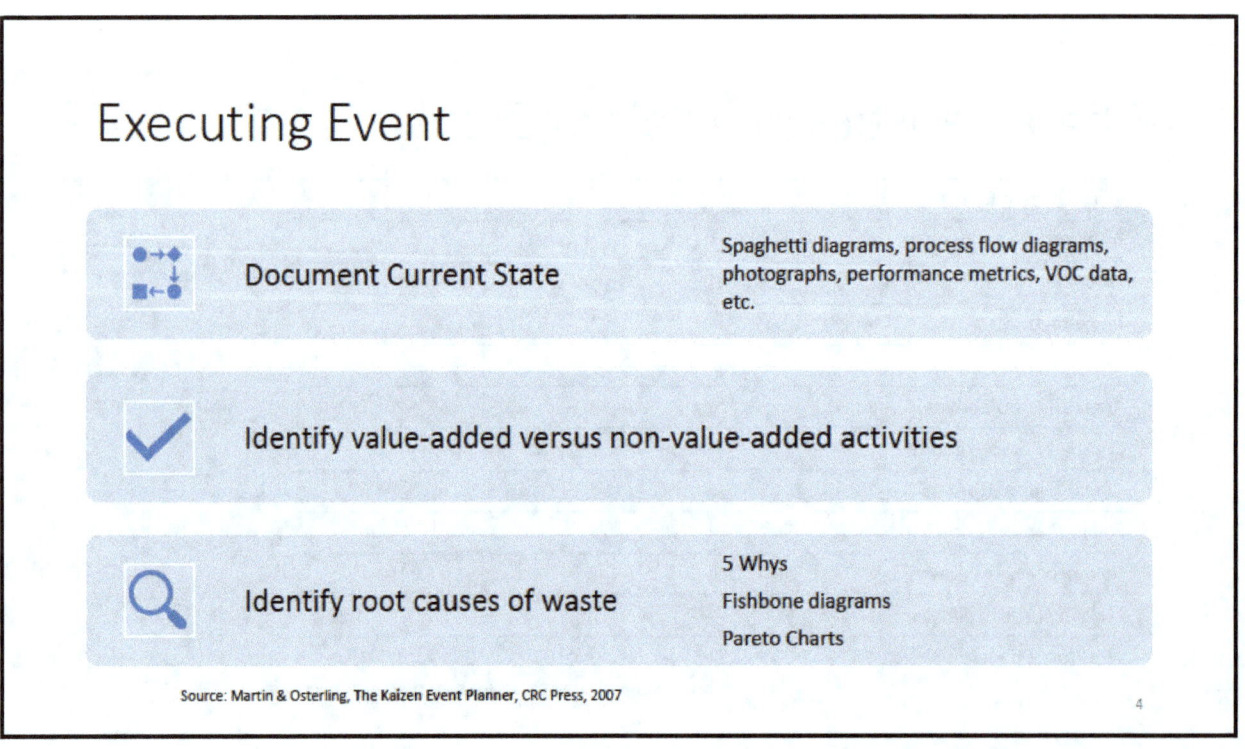

Executing the Event

- Brainstorm and prioritize improvement options using a Priority-Action-Consider-Eliminate (PACE) Diagram

Source: Martin & Osterling, **The Kaizen Event Planner**, CRC Press, 2007

Executing the Event

- Design, Test & Implement Improvements
- Wrap-Up
    - Kaizen Event Report
    - Sustainability Plan
    - 30-Day List (punch list)
    - Parking Lot Issues
    - Team Presentation
- Team Recognition & Celebrating
- Disband Team

Source: Martin & Osterling, **The Kaizen Event Planner**, CRC Press, 2007

Control Charts

Define → Measure → Analyze → Improve → **Control**

Variation Exercise

Everything varies!
1. Random or common cause variation
2. Special cause variation

Name Game:

Write your name 5 times. Now write your name with your other hand.

Purpose of Control Charts

- A graphical way to track a process variable over time
- A statistical method to determine whether the process has changed
    - Is variation due to common causes?
    - Is variation due to special causes?

Out-of-Control

- Control charts can signal an out-of-control condition
- Indicates that a special cause of variation is present
- Out-of-control implies that the process is not predictable

Recall: How Unusual?

Rule of thumb:

A **Z score** less than -2 or greater than +2 is considered unusual

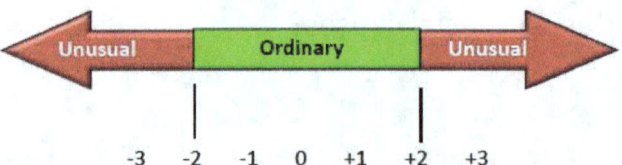

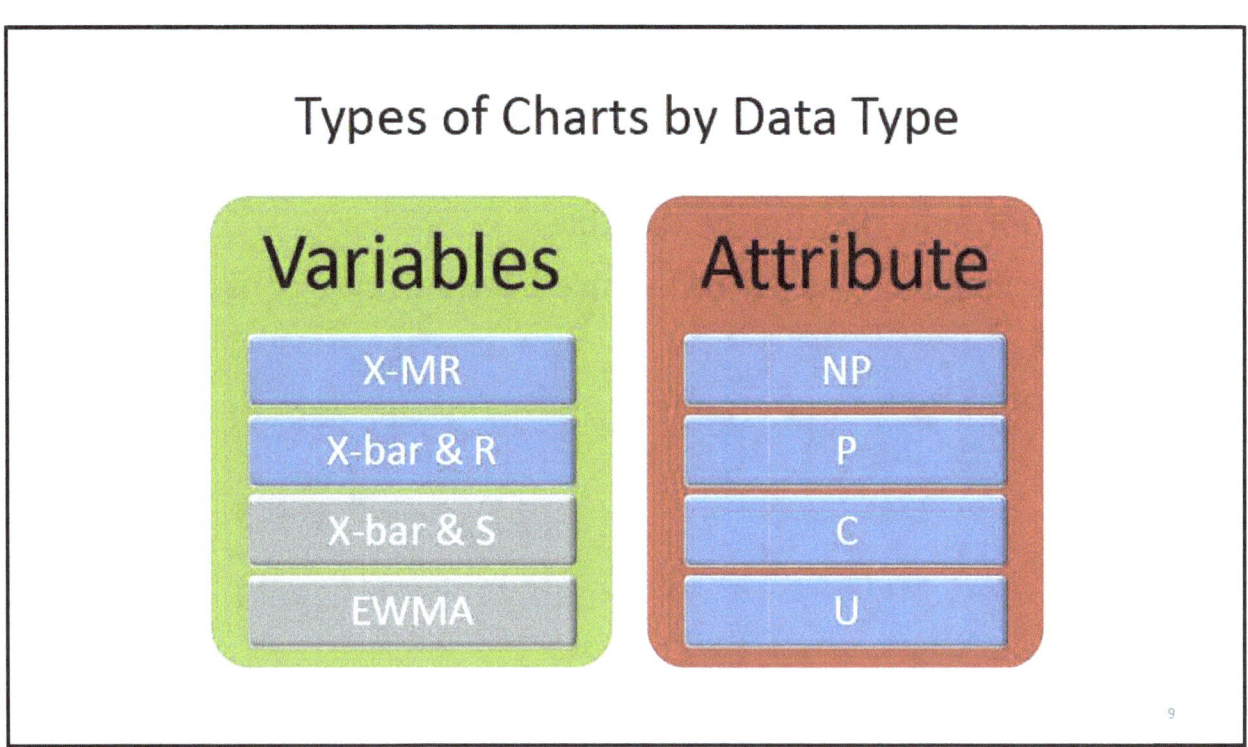

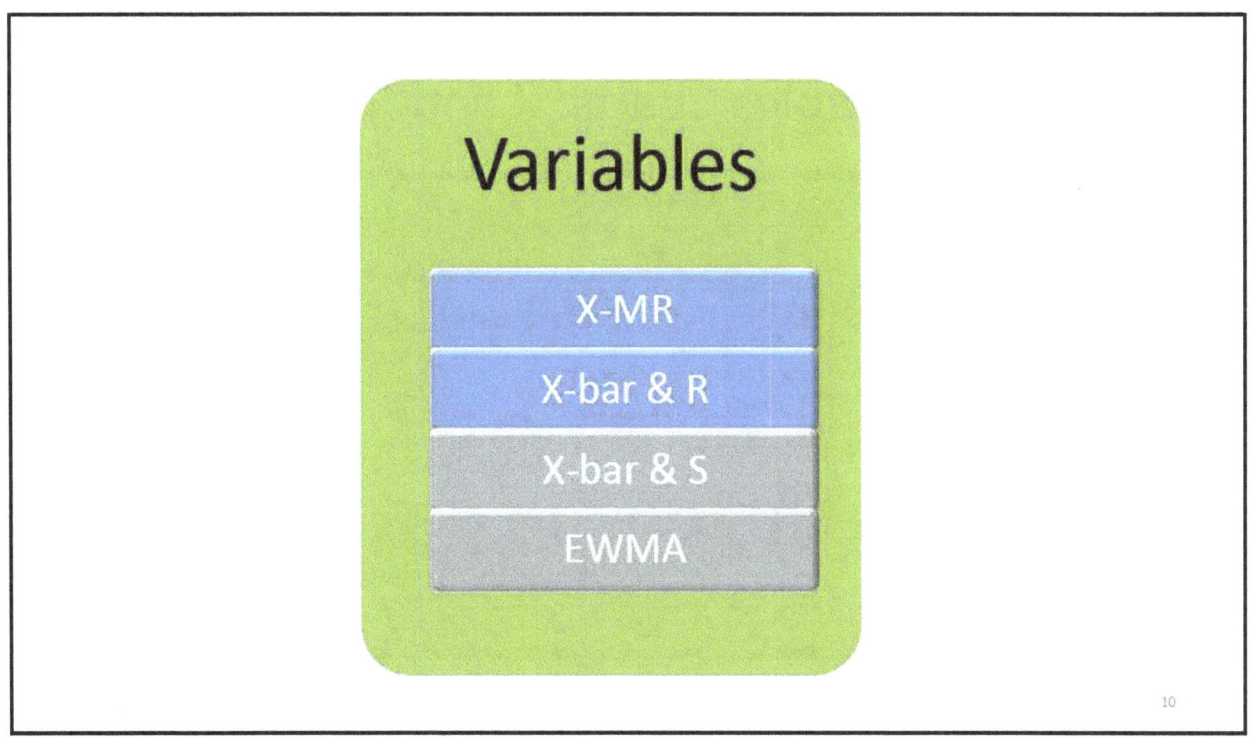

Variables Charts: X and MR

- **X is a normally distributed variable**
- **The moving range is the difference between two consecutive points**
- **Subgroup size = 1**

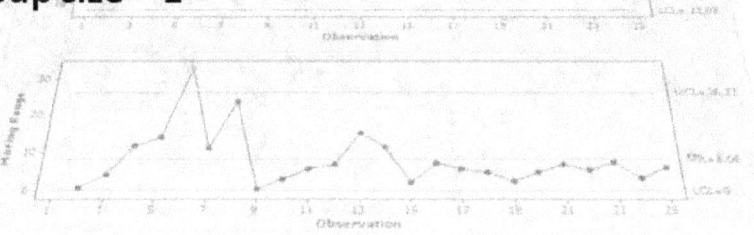

Variables Charts: X and MR

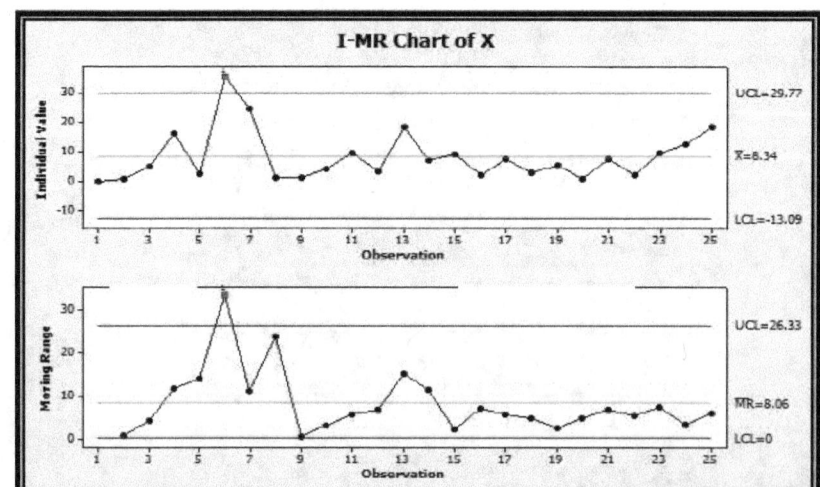

Variables Charts: X-Bar and R Chart

- **Use subgroup sizes from 3 to 6**
- **X-bar is the average of the subgroup**
  - Tracks the location of the process
- **R is the range of the subgroup**
  - Tracks the variability within the subgroups

Variables Charts: X-bar and R

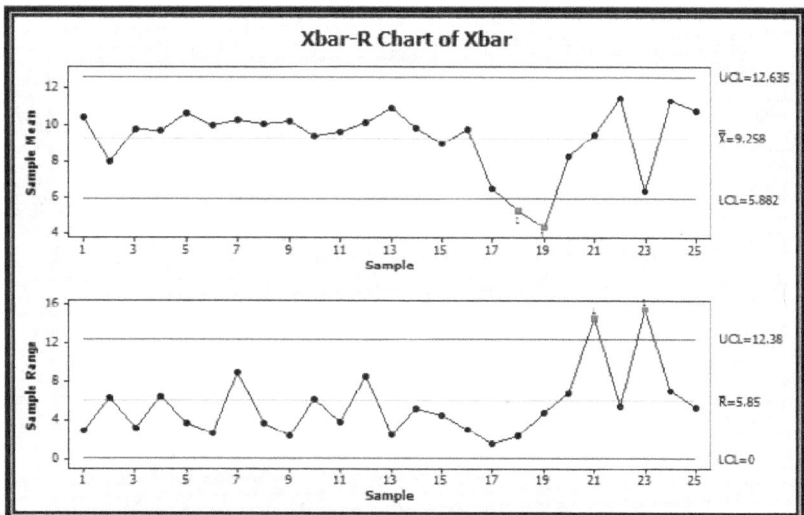

Attribute

- NP
- P
- C
- U

Defectives vs. Defects

- A **defective** unit is found using a pass/fail, go/no go type of test

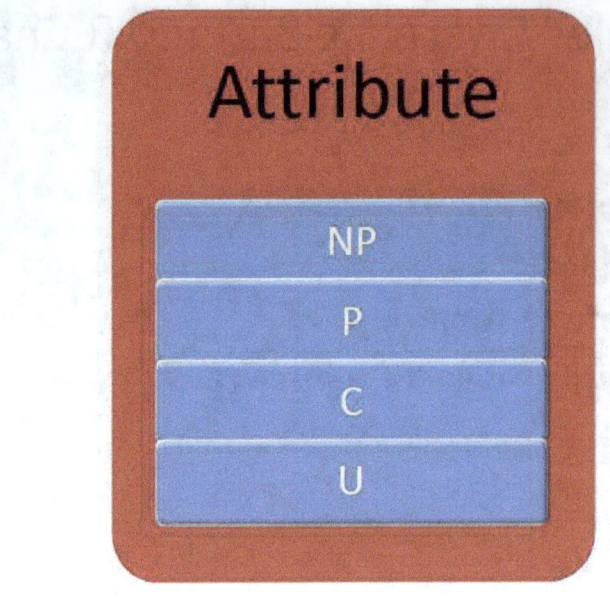

- There can be many **defects** in a single unit

Attributes Charts: NP

- Products are classified as **defective** or non-defective (go/ no go)
- NP is the number of **defective** units found in the sample
- Sample size is constant

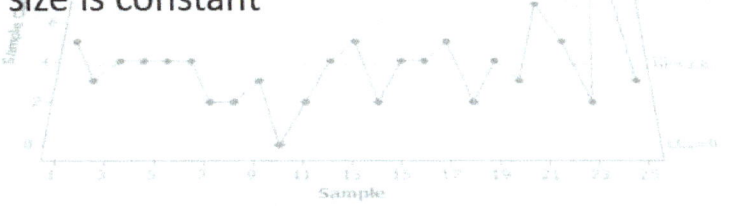

Attributes Charts: NP

NP Chart of p

UCL=9.54
$\overline{NP}$=3.8
LCL=0

Attributes Charts: P

- Products are classified as **defective** or non-defective (go/no go)
- P is the percent **defective** in the sample
- Sample size may not be constant
- Control limits change width according to sample size

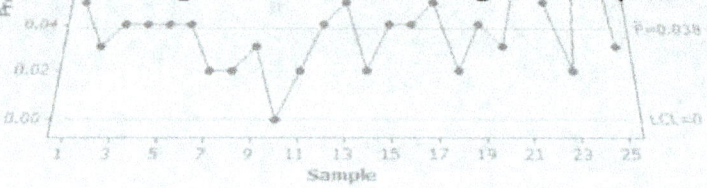

Attributes Charts: P

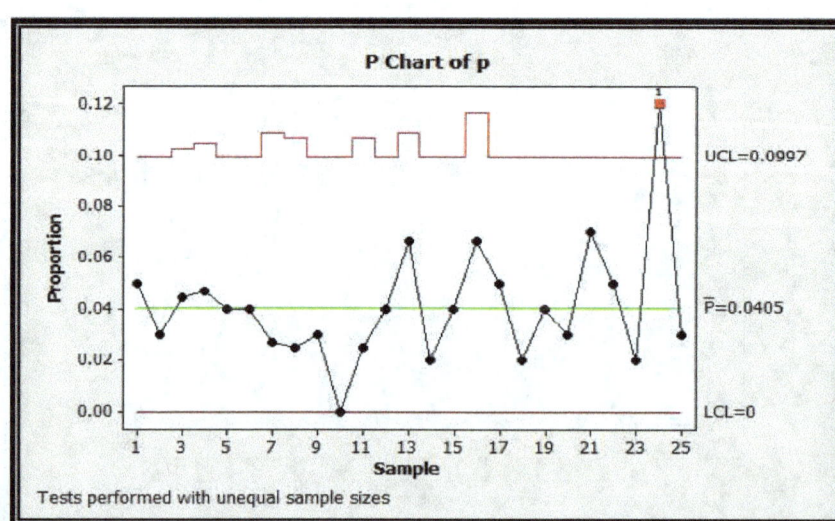

Tests performed with unequal sample sizes

Attributes Charts: C

- Products are examined and the number of **defects** is recorded
- C is the number of **defects** found in a sample
- Sample size is constant

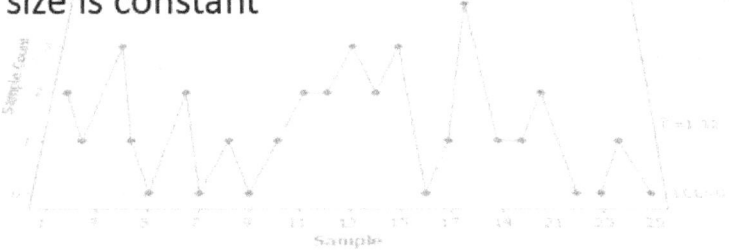

Attributes Charts: C

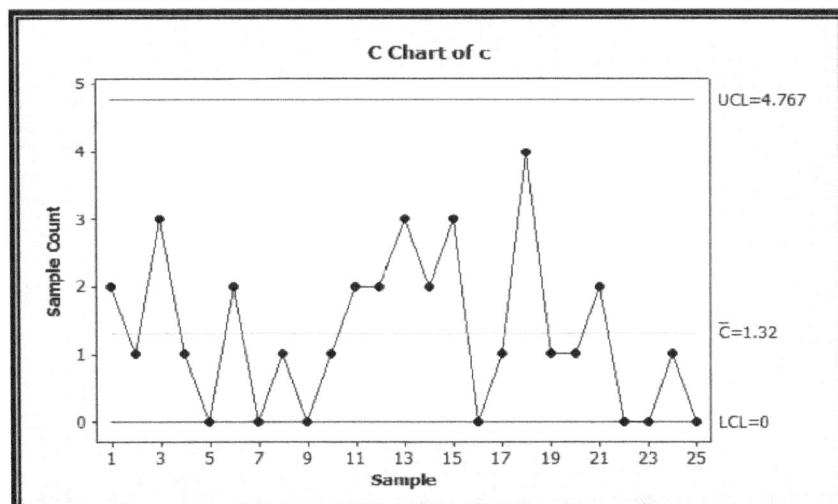

Attributes Charts: U

- Products are examined and the number of **defects** is recorded
- U is the average number of **defects** per unit
- Sample size may not be constant
- Control limits change width according to sample size

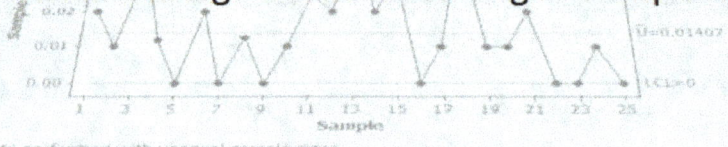

Attributes Charts: U

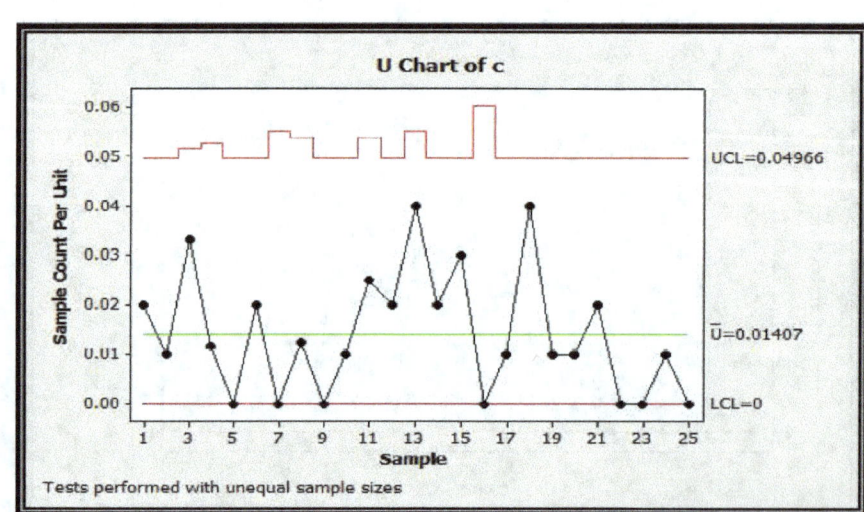

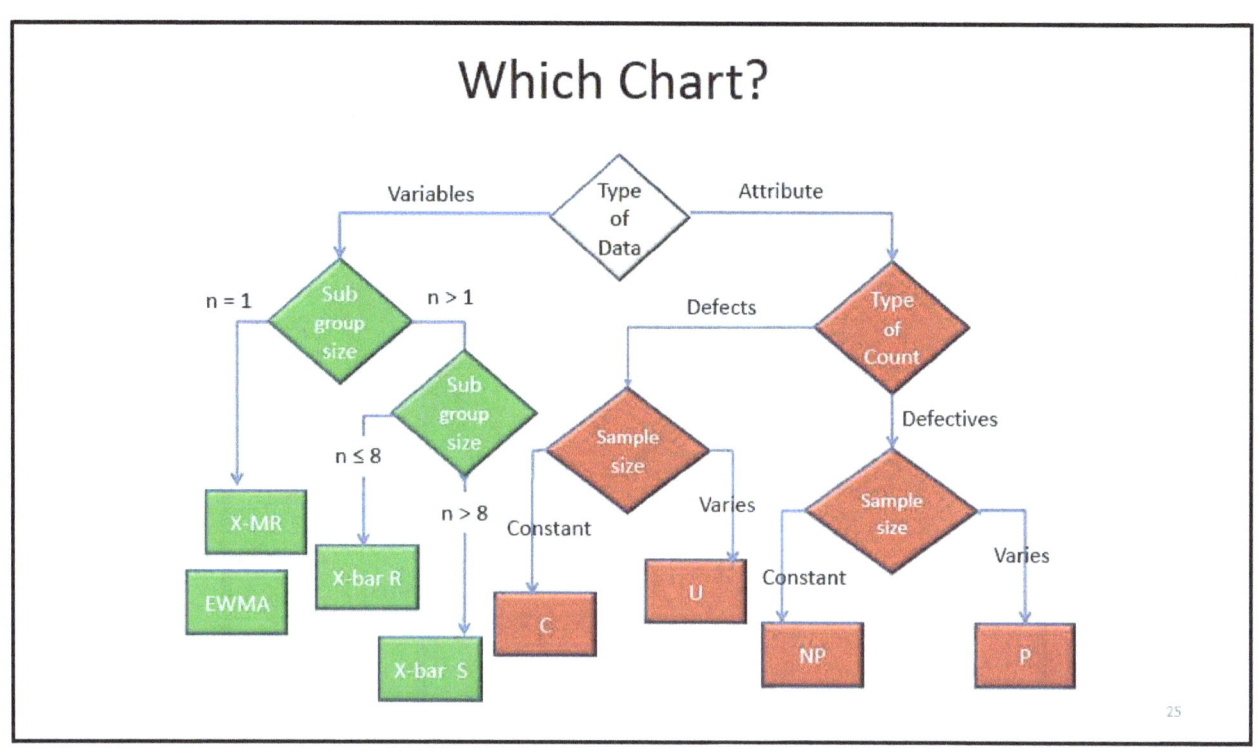

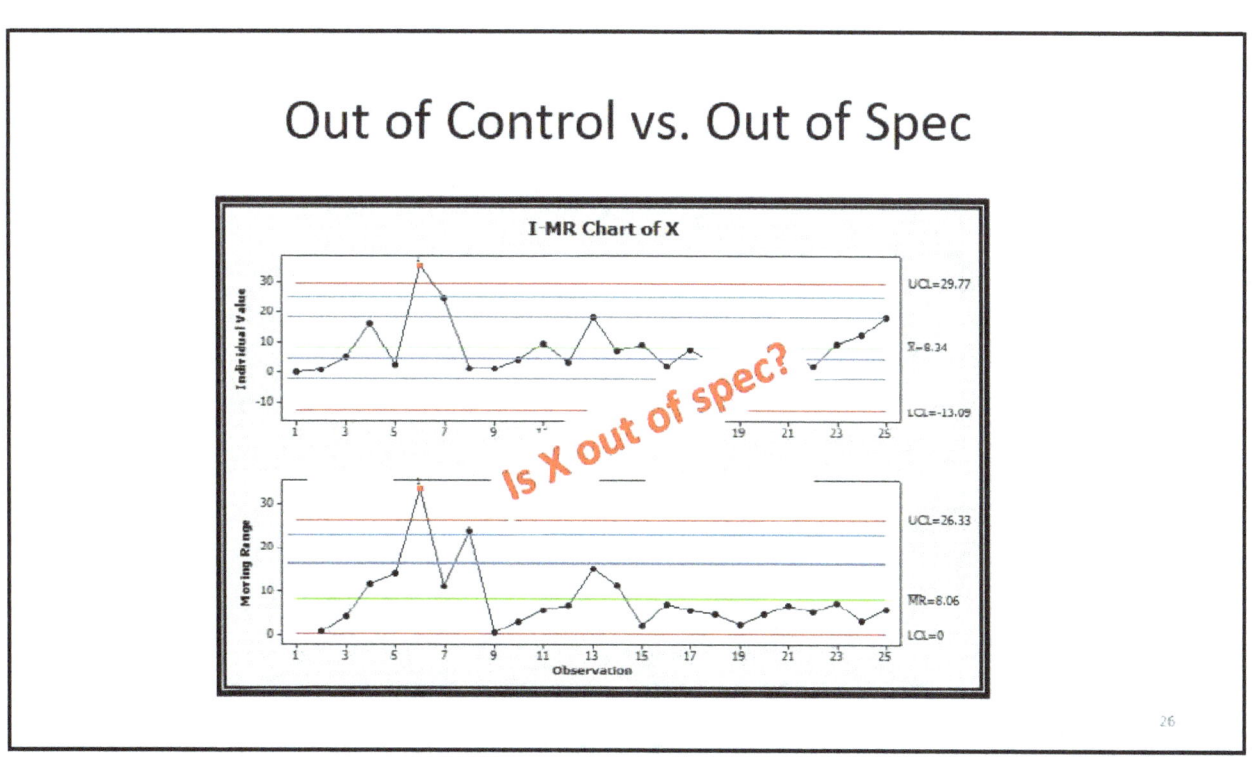

NP Chart: Out of Spec?

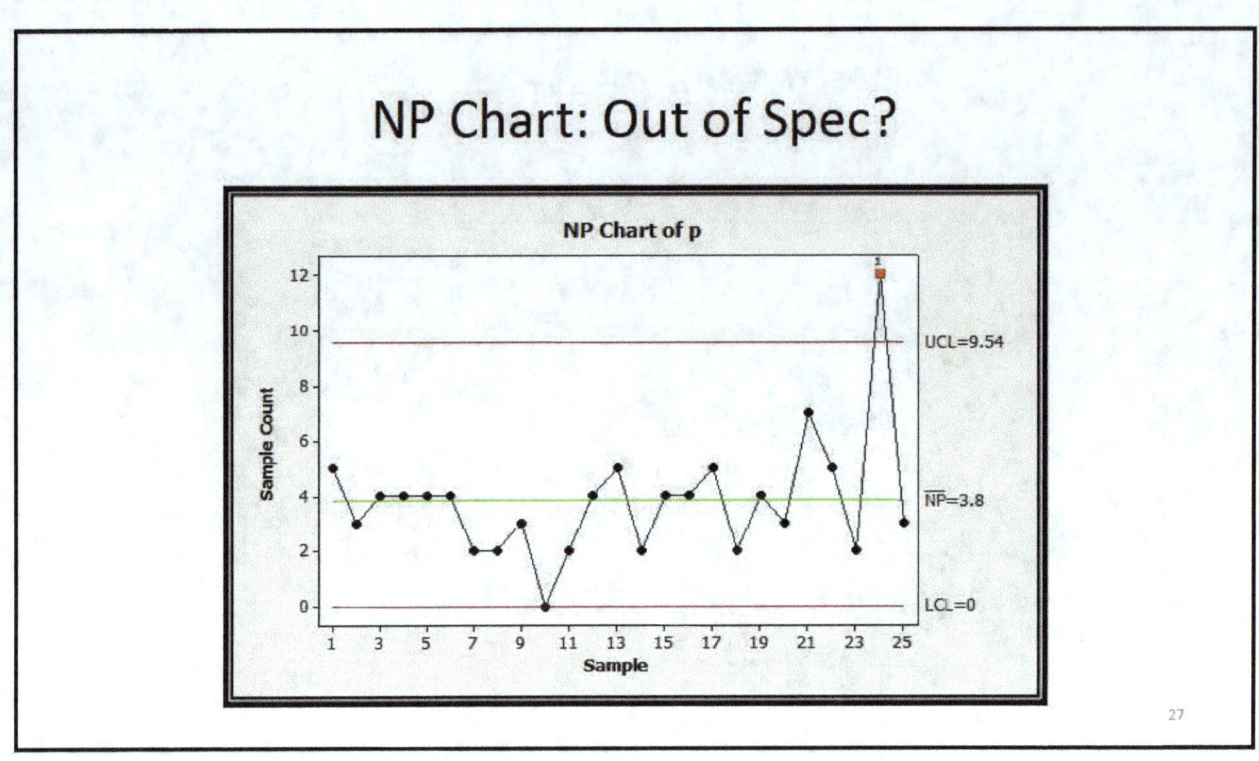

C Chart: Out of Spec?

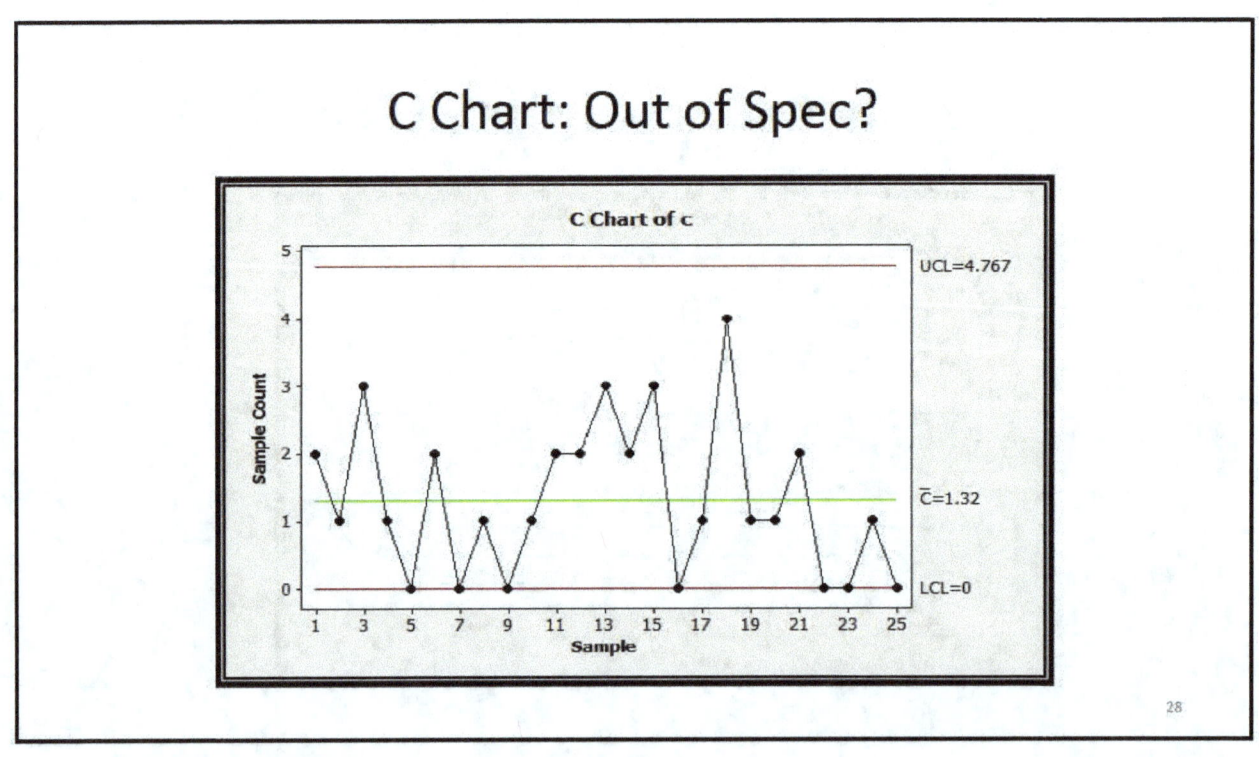

Which Chart?

Types of Charts by Data Type

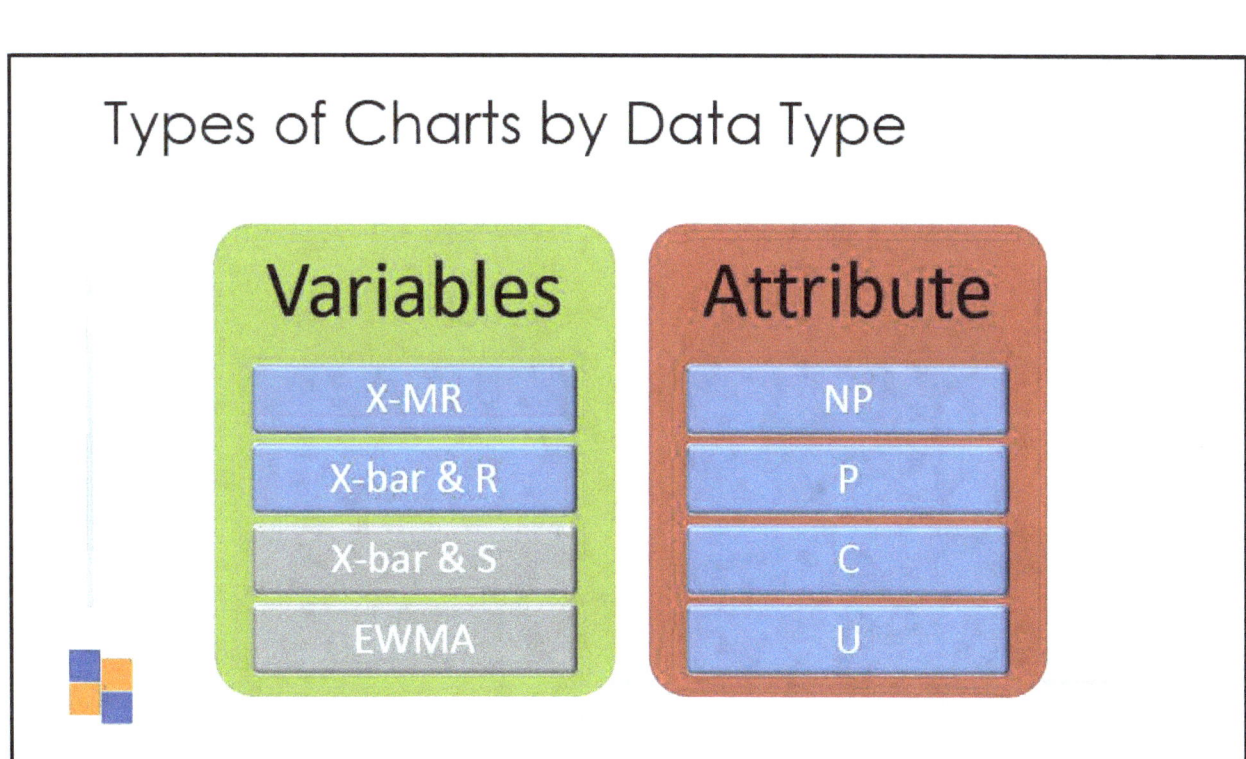

Defectives vs. Defects

- A **defective** unit is found using a pass/fail, go/no go type of test

- There can be many **defects** in a single unit

Which Chart?

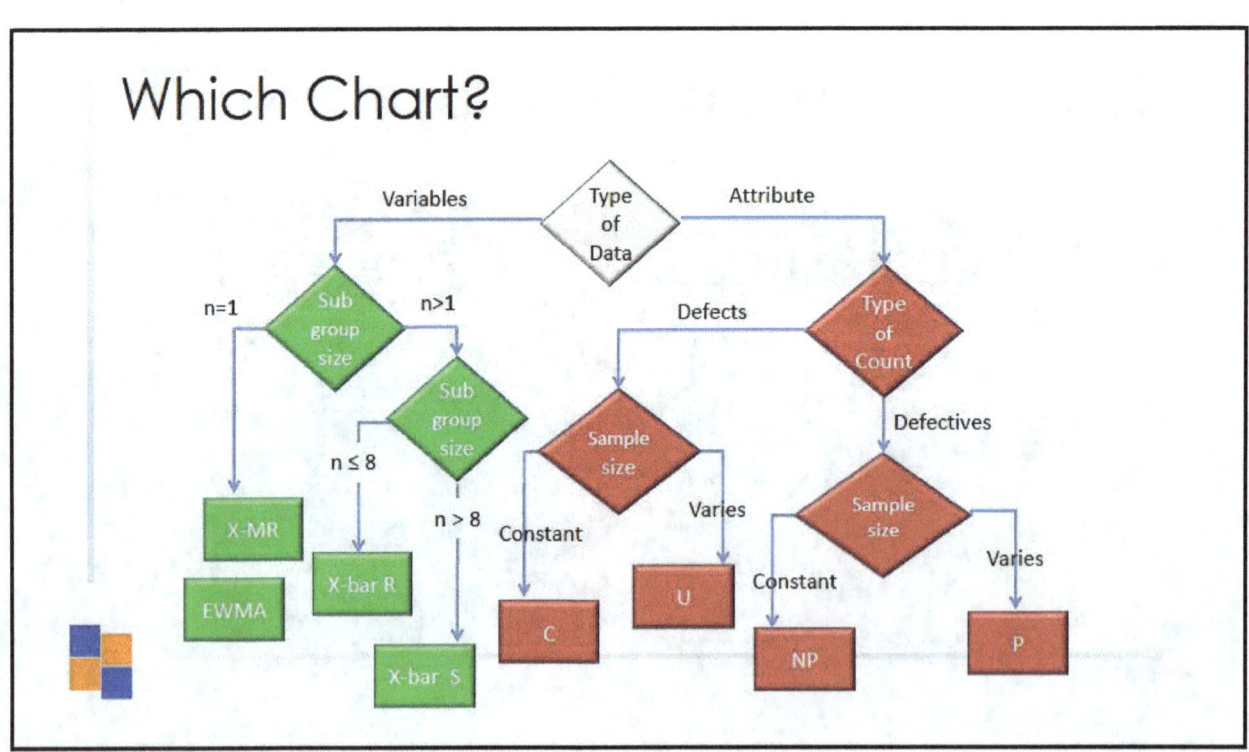

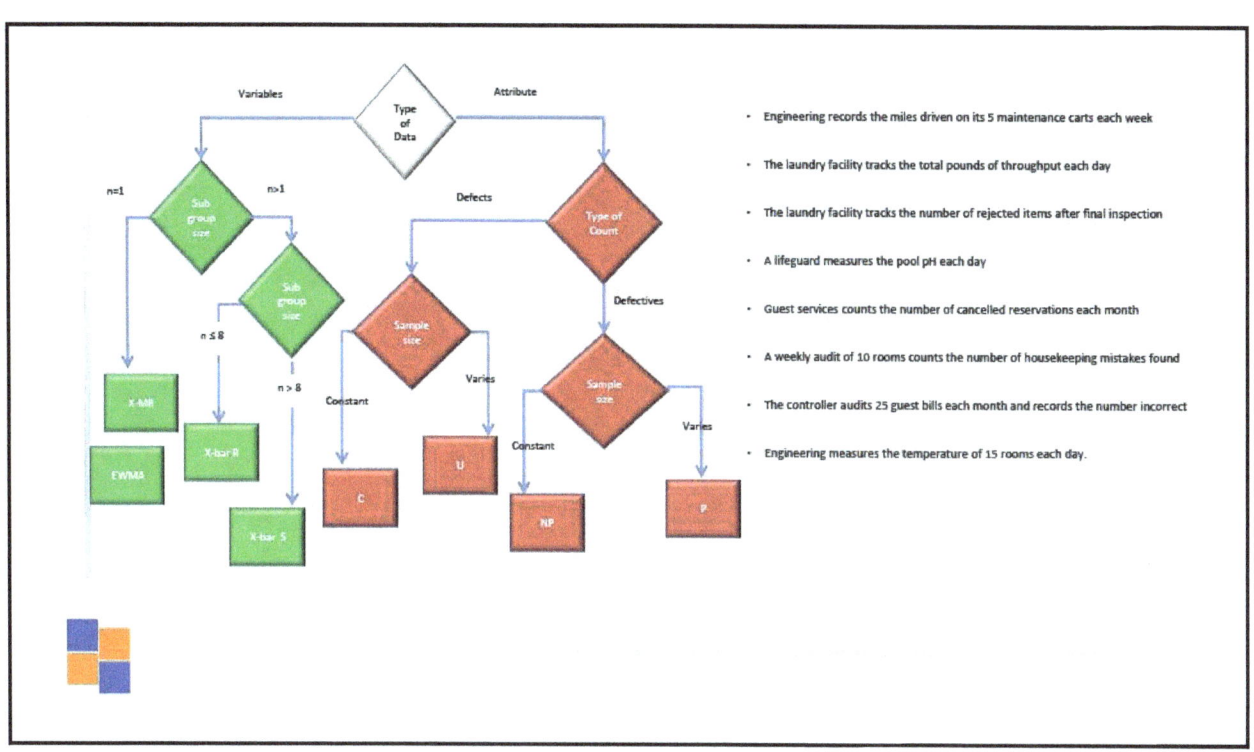

Control Plans

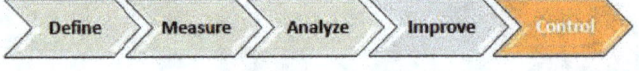

Control Plans

- Document process changes through SOPs
- Monitor key inputs and outputs
  - Control charts
  - Check lists
  - Mistake proofing
  - Visual management
- Create a corrective action plan to address anomalies

Closing the Project

- Hand control back to the process owners

- Celebrate success

- Share lessons learned with the organization

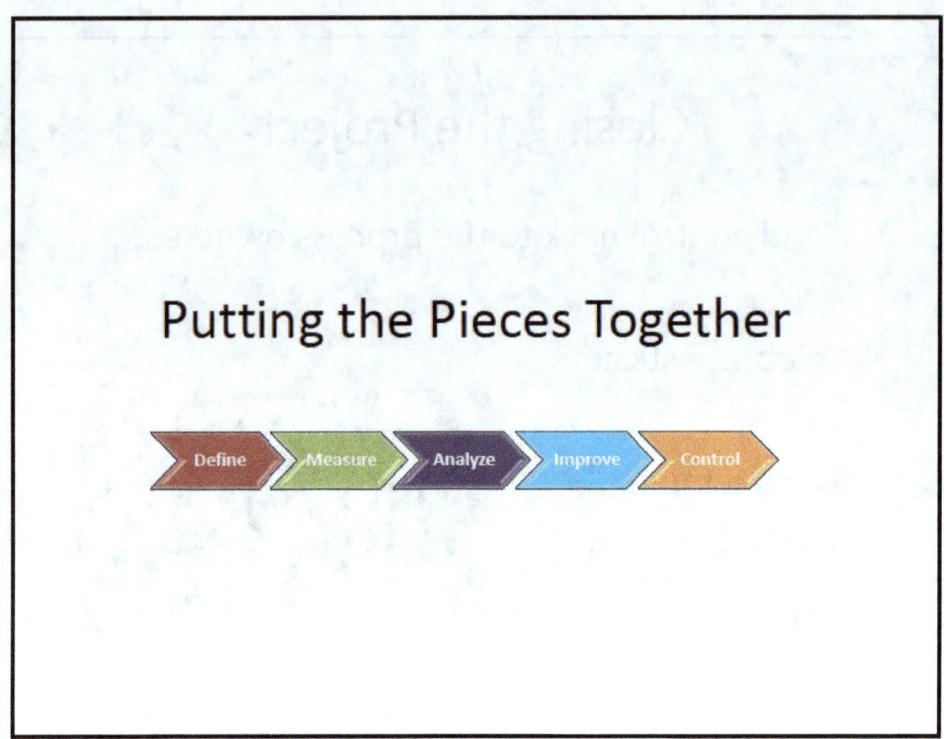

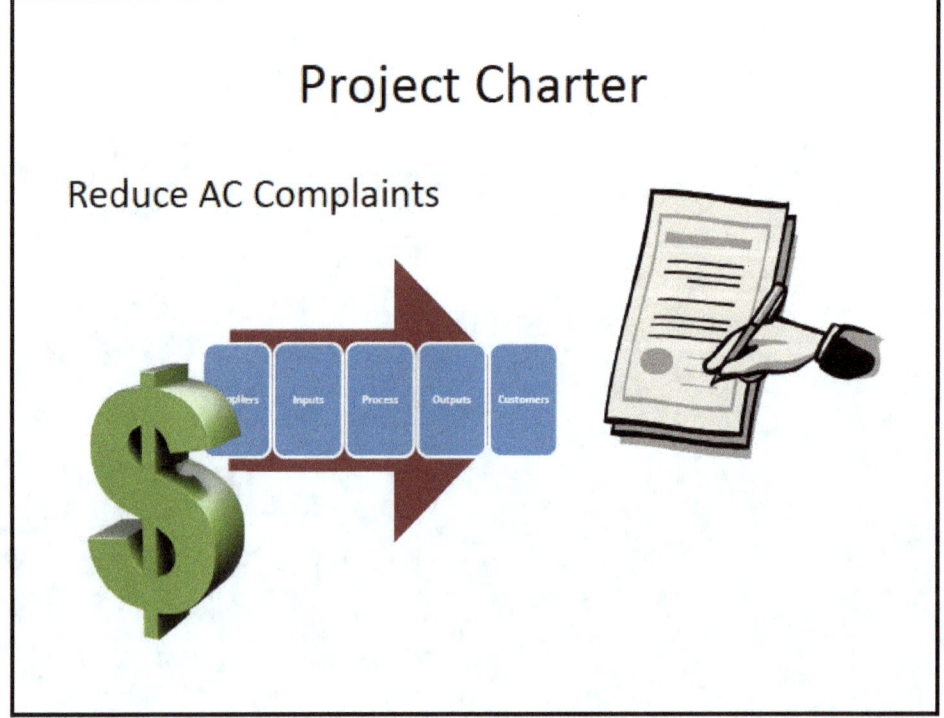

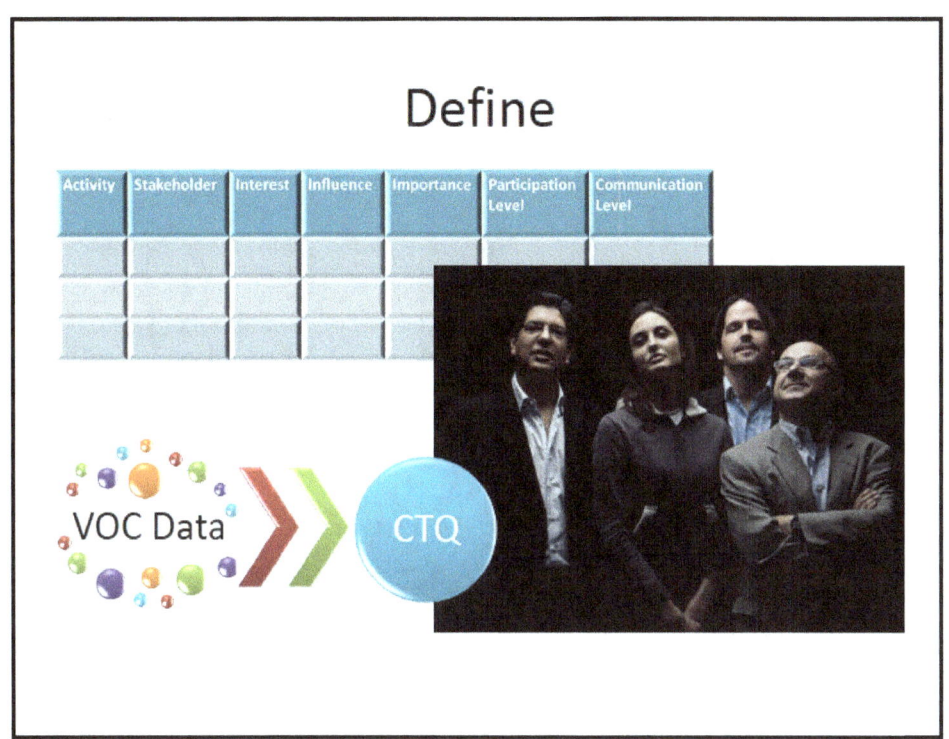

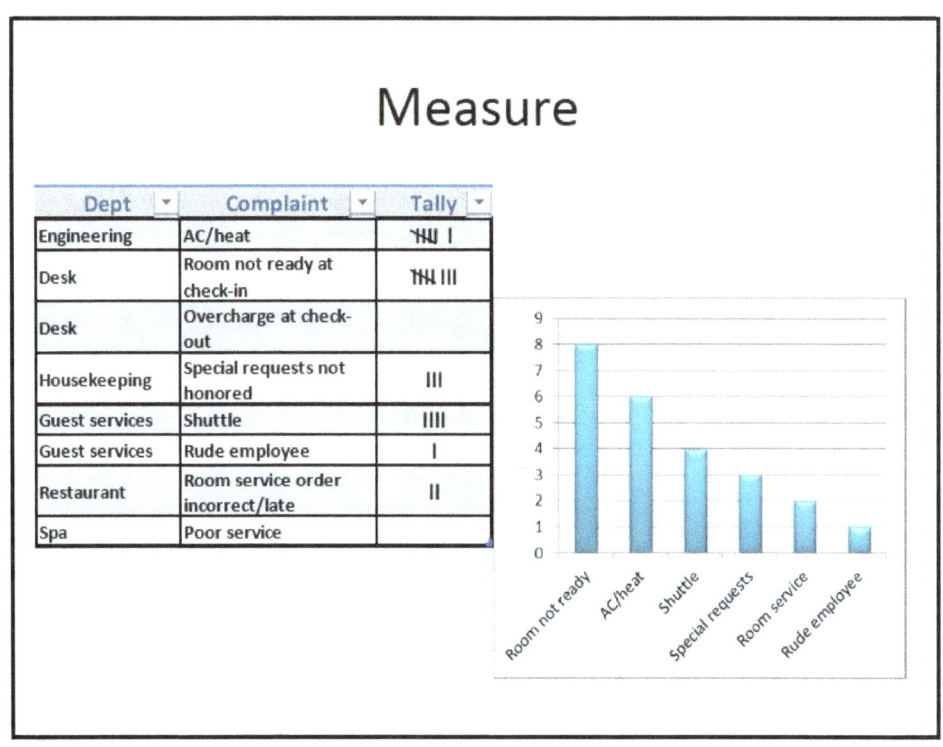

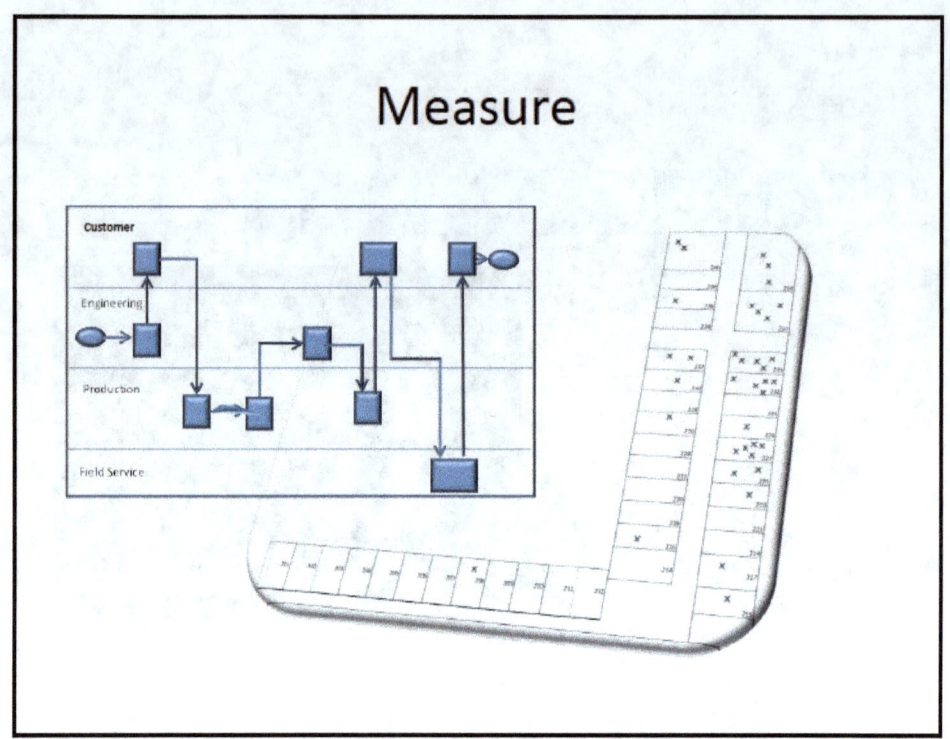

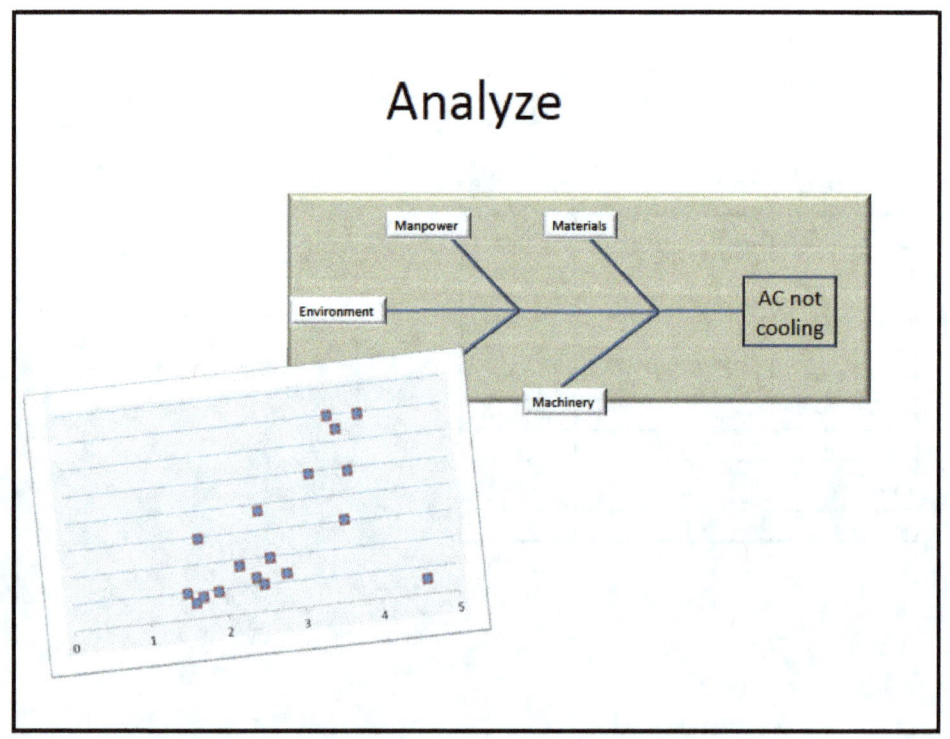

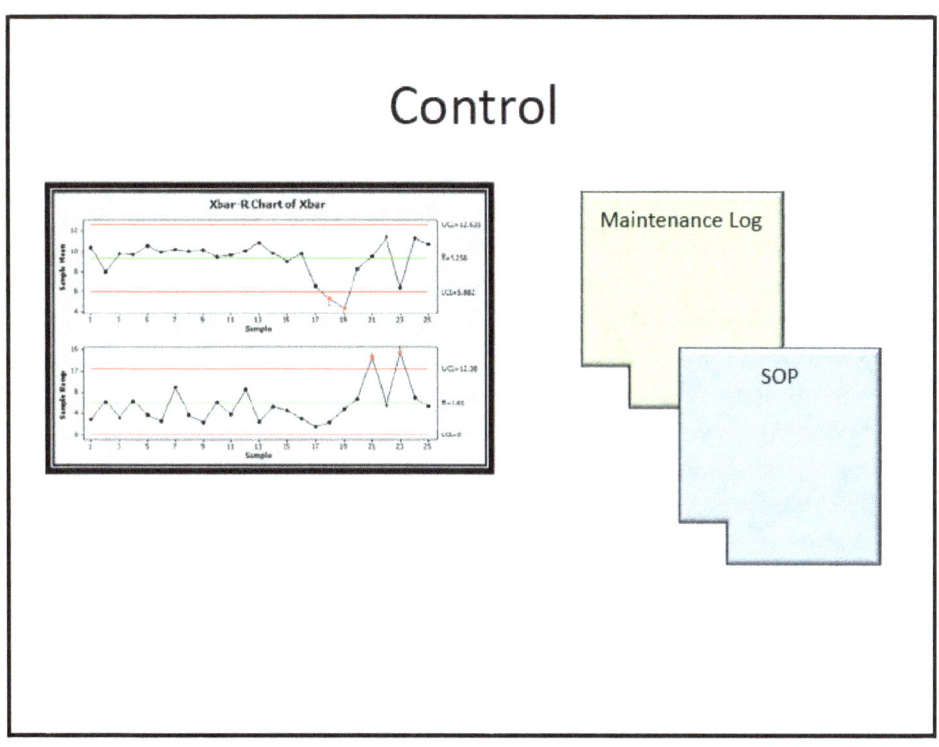

Project Completion

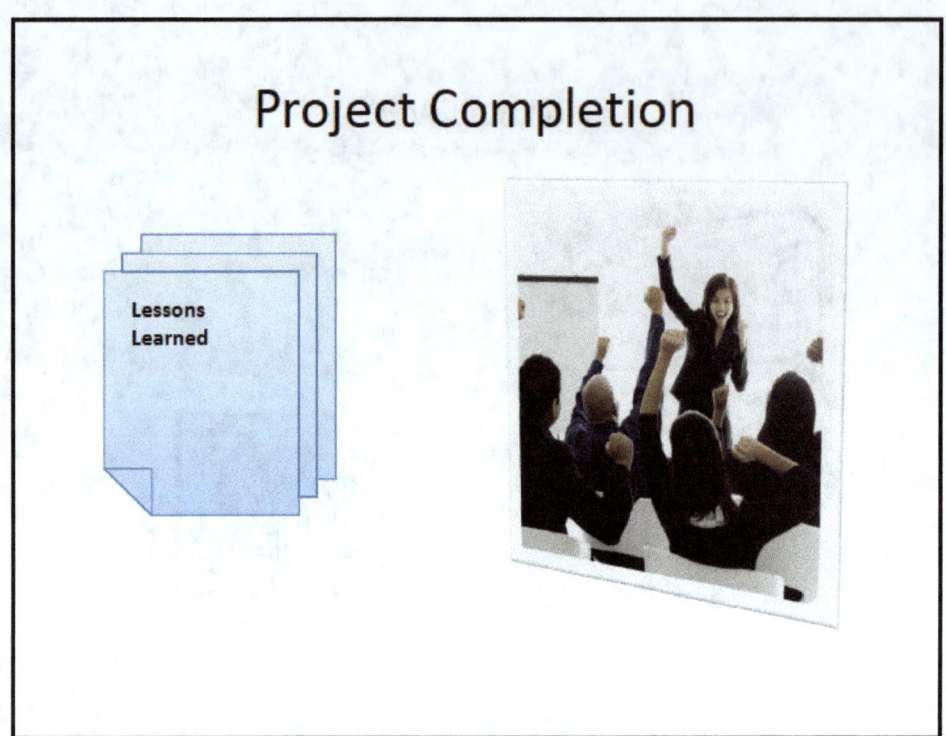

Mirasol Project Reports

Jim Sutton – Project Lead

AC Complaints Project

Six Sigma Project Summary				
Project Title	**Scope (location, area, line, department, process boundaries)**			
Reduce AC Complaints	Maintenance and housekeeping departments, contractor			
Process Name	**Start Date**		**Expected End Date**	
AC system	3/15/20xx		6/1/20xx	
Problem Statement	**Expected Benefits**			
Over the last six months of 2011, AC complaints have increased to 3.3% of turnovers resulting in unhappy guests and room change requests.	Reducing AC complaints will increase customer satisfaction and improve percentage of returning guests. Eliminating AC complaints will save $3500 in comped meals & spa treatments, and room upgrades per year.			
Goal Statement	**Key Metrics**			
By June, address AC problems and eliminate AC complaints.	Measure	Units	Baseline	Goal	Goal Met?
	Complaints	Count/turnvoer	3.30%	0%	Yes
	Temp	F	73	67	68/No
Team Members				
Name	Role	Name		Role
Jim Sutton	Lead	Joe McGill		Member
Jon Sand	Champion			
Ed Durst	Member			

AC Complaints Project

Measles Chart	Root Causes Identified
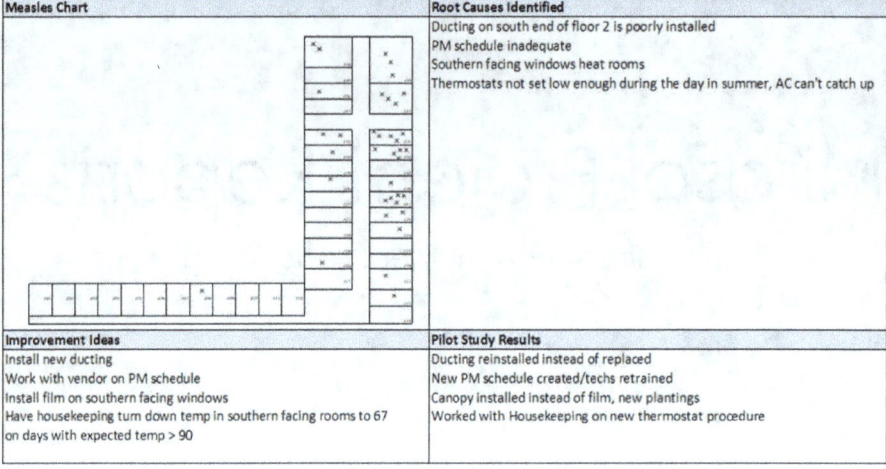	Ducting on south end of floor 2 is poorly installed PM schedule inadequate Southern facing windows heat rooms Thermostats not set low enough during the day in summer, AC can't catch up
Improvement Ideas	**Pilot Study Results**
Install new ducting Work with vendor on PM schedule Install film on southern facing windows Have housekeeping turn down temp in southern facing rooms to 67 on days with expected temp > 90	Ducting reinstalled instead of replaced New PM schedule created/techs retrained Canopy installed instead of film, new plantings Worked with Housekeeping on new thermostat procedure

---

AC Complaints Project

Verification of Benefits	Lessons Learned
Southern facing room temps have been holding at 68 degrees on hottest days No AC complaints in May or June	Mechanical problems don't necessarily have to cost a lot of $ to fix Repair before replacing Many little improvements can add up The measles chart was the key in finding the causes of the problem

Control Plan				Signatures	
Key Metrics	Target	How Monitored	Responsible Party	Champion	_____
AC complaints/turnover	0	Weekly dashboard	Jim Sutton	Leader	_____
Room temps	67	X-bar R chart	Joe McGill	Process Owner	_____
Housekeeping % compliance	100	Check sheet	Nina Sanchez		
PM % compliance	100	Check sheet	Ed Durst	Close Date	7/1/20xx

Golf Cart Availability Project

Project Title		Scope (location, area, line, department, process boundaries)			
Increase golf cart availability		Engineering department			
Process Name		**Start Date**		**Expected End Date**	
Engineering garage		3/15/20xx		7/15/20xx	
Problem Statement		**Expected Benefits**			
During June - August of 20xx, there were a total of 12 complaints about late shuttles (1.7% of turnovers), resulting in unhappy guests and comped meals and spa treatments.		Inceasing golf cart availability wil lincrease customer satisfaction and reduce customer complaints about island transportation. Solving the availability problem will save $1000 in comped meals and spa treatments.			
Goal Statement		**Key Metrics**			
Increase golf cart availability to 95% and elimiante shuttle complaints within 4 months.		Measure	Units	Baseline	Goal	Goal Met?
		Complaints	Count/changeover	1.70%	0	
		Availability	%	80	95	Yes
Team Members					
Name	Role	Name		Role	
Jim Sutton	Lead/YB	Ron Ridpath		Member	
Jon Sand	Champion/GB	Tully Franklin		Member	
Sam Napier	Member				

Golf Cart Availability Project

5S Before and After Photos	Root Causes Identified
	Not enough room for required parts
	Supplier is unreliable/slow
	PM schedule not correct
	Carts are old
	Disorganized inventory system
Improvement Ideas	**Pilot Study/ Improvement Actions**
Perform a 5S on the cart garage	5S created 10% free space, found $2500 in "missing" replacement parts
Identify new parts suppliers	Two new suppliers identified, one under contract
Rework PM schedule	PM rotation schedule reworked, and new assignment board created
Develop a cart replacement schedule	Cart replacement schedule
Computerize parts inventory	Bar code and computer system deemed too much for our small shop; worked with intern to develop an Access database system inhouse

Golf Cart Availability Project

Verification of Benefits
Parts now at recommended levels
One cart taken out of service, replacement ordered
Carts now operating at 98% availability
No complaints for last 2 months

Lessons Learned
Keeping an organized garage is key
Don't be afraid to find new vendors if needed
Buying your way out of a problem might not be necessary
Lean tools can provide quick and effective improvements
A Six Sigma project doesn't have to be intimidating
Remember to take before and after pictures when doing a 5S

Control Plan

Key Metrics	Target	How Monitored	Responsible Party
% PM compliance	100	Weekly dashboard	Ron Ridpath
Parts inventory	Varies by part	Weekly database report	Sam Napier
Supplier on-time delivery	100%	Monthly report	Tully Franklin
Availability	100%	P chart	Ron Ridpath
Complaints	0%	Weekly report	Jim Sutton

Signatures

Champion _____

Leader _____

Process Owner _____

Close Date

Mirasol Project Reports

Eli Guzman – Project Lead

Laundry Water Project

Six Sigma Project Summary						
Project Title		**Scope (location, area, line, department, process boundaries)**				
Decrease laundry water usage		Housekeeping department, laundry team, contractor				
Process Name		**Start Date**			**Expected End Date**	
Commercial laundry facility		1/15/20xx			9/15/20xx	
Problem Statement		**Expected Benefits**				
Water usage in the commercial laundry facility has increased by 25% year over year, resulting in an additional $15,000 in expenses.		Savings due to reduction in water usage to previous levels: Cost to repair/replace system: XXXXX First year net $ savings: XXXXX Subsequent years net savings: XXXXXX				
Goal Statement		**Key Metrics**				
Reduce laundry water usage to previous levels within 8 months.		Measure	Units	Baseline	Goal	Goal Met?
			Efficiency	Gal/lb	5.0	4.0	2.5/Yes
			Rework	%	5	5	Yes
			Cycle time	Min/lb	0.5	0.5	0.48/Yes
			Cpk	NA	0.111	1.67	1.55/No
Team Members							
Name	Role		Name			Role	
Eli Guzman	Lead		Joe Ballard			Member	
Fred Sand	Champion						
Veronica LeMay	Member						

Laundry Water Project

Histogram	Root Causes Identified
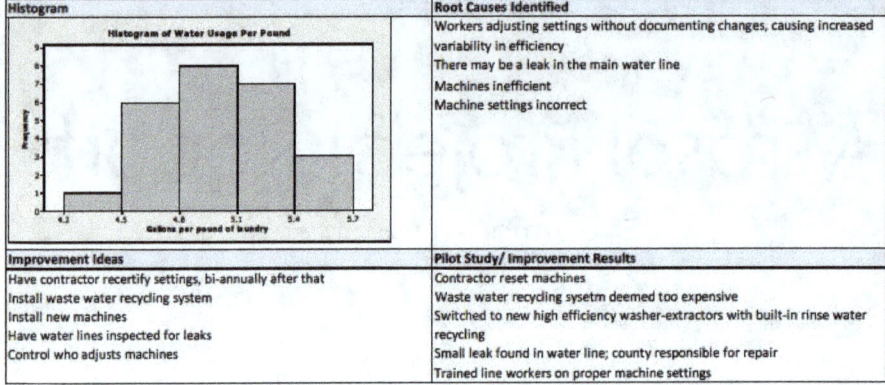	Workers adjusting settings without documenting changes, causing increased variability in efficiency There may be a leak in the main water line Machines inefficient Machine settings incorrect
Improvement Ideas	**Pilot Study/ Improvement Results**
Have contractor recertify settings, bi-annually after that Install waste water recycling system Install new machines Have water lines inspected for leaks Control who adjusts machines	Contractor reset machines Waste water recycling sysetm deemed too expensive Switched to new high efficiency washer-extractors with built-in rinse water recycling Small leak found in water line; county responsible for repair Trained line workers on proper machine settings

Laundry Water Project

Verification of Benefits	Lessons Learned
County has repaired leak in water line New machines have been running at 2.5 gal/lb, std dev = 0.1 gal Rework percentage has not increased. New machines have not increased cycle time July net savings: 35% ($ figure redacted) Expected annual net savings: 40% ($ figure redacted)	The entire department was involved in planning the washer install which made the transition run very smoothly Laundry line workers are now responsible for monitoring machine settings Workers are now trained to set machines themselves

Control Plan

Key Metrics	Target	How Monitored	Responsible Party
Efficiency	2.5 gal/lb	X-MR chart	Joe Ballard
Rework	10%	P Chart	Veronica LeMay
Machine settings	Varies	Daily check sheet	Line workers
Cpk	1.67	Via X-MR chart	Eli Guzman

Signatures

Champion _____
Leader _____
Process Owner _____

Close Date   9/7/20xx

Rooms Not Ready Project

Six Sigma Project Summary					
Project Title	**Scope (location, area, line, department, process boundaries)**				
Reduce complaints for "rooms not available at check in"	Housekeeping and desk				
Process Name	**Start Date**		**Expected End Date**		
Housekeeping room turnover	4/15/20xx		7/15/20xx		
Problem Statement	**Expected Benefits**				
In the past year, Mirasol has received 36 complaints about rooms not being ready at check-in (with a mean shift in July) resulting in unhappy guests.	Increased turnover efficiency will reduce customer complaints at check in and save $2000 annually in comped meals/spa treatments while new guests wait. Increase efficiency will result in a cost avoidance of hiring a new housekeeper in high season at $11/hour.				
Goal Statement	**Key Metrics**				
Eliminate "room not ready at check-in" complaints within 2 months.	Measure	Units	Baseline	Goal	Goal Met?
	Room not ready at check-in complaints	Count/turnover	3.50%	0%	Yes
	Room cleanliness complaints	Count/turnover	0	0	Yes
Team Members					
Name		**Role**	**Name**		**Role**
Eli Guzman		Lead	Nina Sanchez		Member
Fred Sand		Champion	Cindy Sparks		Member
Dave Harris		Member			

Rooms Not Ready Project

Histogram	Root Causes Identified
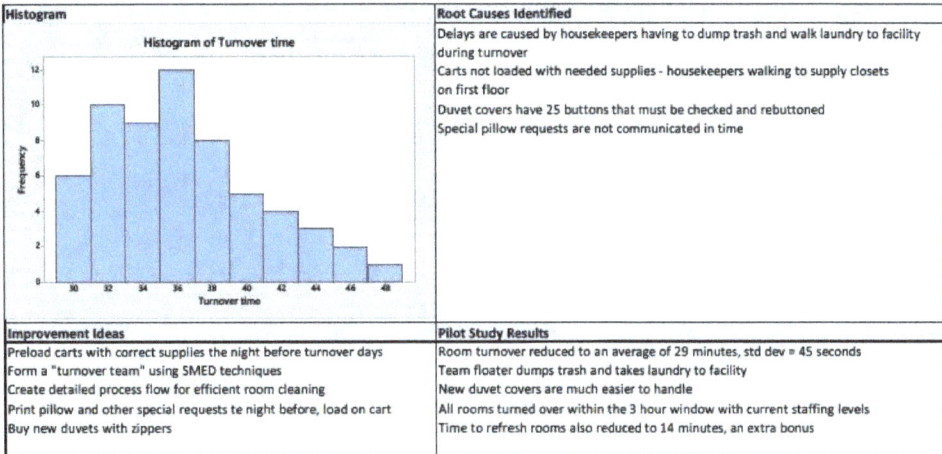	Delays are caused by housekeepers having to dump trash and walk laundry to facility during turnover Carts not loaded with needed supplies - housekeepers walking to supply closets on first floor Duvet covers have 25 buttons that must be checked and rebuttoned Special pillow requests are not communicated in time
Improvement Ideas	**Pilot Study Results**
Preload carts with correct supplies the night before turnover days Form a "turnover team" using SMED techniques Create detailed process flow for efficient room cleaning Print pillow and other special requests te night before, load on cart Buy new duvets with zippers	Room turnover reduced to an average of 29 minutes, std dev = 45 seconds Team floater dumps trash and takes laundry to facility New duvet covers are much easier to handle All rooms turned over within the 3 hour window with current staffing levels Time to refresh rooms also reduced to 14 minutes, an extra bonus

Rooms Not Ready Project

Verification of Benefits	Lessons Learned
New system used on three high turnover Saturdays in the high season - all rooms ready for guests by check in	Tools used: spaghetti diagram, process flow chart, SMED
No additioanl staff required	The housekeeping staff was involved in developing new procedures and
No reduction in room cleanliness	brainstormed ways to reduce errors. This made the group accept the changes
All housekeepers trained on new procedures	in their workflow, and improved housekeeping morale.
Attendance increased from 85% to 90% on turnover days	Housekeeping attendance has improved, but more needs to be done

Control Plan				Signatures	
Key Metrics	Target	How Monitored	Responsible Party	Champion	_____
Turnaround time	29 min	EWMA chart	Eli Guzman	Leader	_____
Attendance	100%	P Chart	Nina Sanchez	Process Owner	_____
Room complaints	0	Run chart	Eli Guzman	Close Date	7/28/20xx

www.ingramcontent.com/pod-product-compliance
Lightning Source LLC
Chambersburg PA
CBHW062215220526
45471CB00009B/3212